Springer Proceedings in Mathematics & Statistics

Volume 166

Springer Proceedings in Mathematics & Statistics

This book series features volumes composed of selected contributions from workshops and conferences in all areas of current research in mathematics and statistics, including OR and optimization. In addition to an overall evaluation of the interest, scientific quality, and timeliness of each proposal at the hands of the publisher, individual contributions are all refereed to the high quality standards of leading journals in the feld. Thus, this series provides the research community with well-edited, authoritative reports on developments in the most exciting areas of mathematical and statistical research today.

More information about this series at http://www.springer.com/series/10533

Yasumasa Nishiura • Motoko Kotani
Editors

Mathematical Challenges in a New Phase of Materials Science

Kyoto, Japan, August 2014

Springer

Editors
Yasumasa Nishiura
Advanced Institute for Materials Research
Tohoku University
Sendai, Miyagi
Japan

Motoko Kotani
Advanced Institute for Materials Research
Tohoku University
Sendai, Miyagi
Japan

ISSN 2194-1009 ISSN 2194-1017 (electronic)
Springer Proceedings in Mathematics & Statistics
ISBN 978-4-431-56102-6 ISBN 978-4-431-56104-0 (eBook)
DOI 10.1007/978-4-431-56104-0

Library of Congress Control Number: 2016946029

Mathematics Subject Classification (2010): 00A71, 34A33, 35B27, 35B36, 35K57, 35Q74, 37Lxx, 49N45, 52C99, 55N99, 74N20, 74R20

Printed on acid-free paper

This Springer imprint is published by Springer Nature
The registered company is Springer Japan KK

Preface

This volume is the proceedings of the RIMS International Conference "Mathematical Challenges in a New Phase of Materials Science", held in Kyoto, August 4–8, 2014, organized by Yasumasa Nishiura, Motoko Kotani, Hisashi Okamoto, Nobuaki Obata, Ken Nakajima, Carlos J. Garcia-Cervera and Thomas P. Russell. This was a satellite conference of Seoul ICM 2014.

The aim of the conference was to provide a forum for presenting and discussing recent trends in mathematical aspects of materials science, and to foster the collaboration between mathematicians and experimentalists multilaterally. The main focuses of the conference were: Non-equilibrium dynamics and dynamical system theory; A Topological approach to nanomaterials: from local to global; Hierarchical materials

More than 60 participants attended the conference and 27 talks were presented. This volume is a collection of the 8 presentations including defect dynamics, negatively curved carbon crystal, topological analysis of di-block copolymers, persistence modules, and fracture dynamics.

The RIMS International Conference would have been impossible without generous financial assistance received from: The Research Institute for Mathematical Sciences (RIMS), Kyoto University; The Japan Science and Technology Agency (JST); Grants-in-Aid for Scientific Research KAKENHI A24244004, Motoko Kotani; Grants-in-Aid for Scientific Research KAKENHI A26247015, Yasumasa Nishiura.

Special thanks are extended to the referees for their valuable comments and to Natsuhiko Yoshinaga for the highly efficient support during the peer-review process, and finally but not least to Masayuki Nakamura of Springer for encouragement and advice.

Sendai, Miyagi, Japan — Yasumasa Nishiura
Sendai, Miyagi, Japan — Motoko Kotani
February 2016

Contents

Generation of Defects and Disorder from Deeply Quenching a Liquid to Form a Solid

A.J. Archer, M.C. Walters, U. Thiele, and E. Knobloch

Abstract We show how deeply quenching a liquid to temperatures where it is linearly unstable and the crystal is the equilibrium phase often produces crystalline structures with defects and disorder. As the solid phase advances into the liquid phase, the modulations in the density distribution created behind the advancing solidification front do not necessarily have a wavelength that is the same as the equilibrium crystal lattice spacing. This is because in a deep enough quench the front propagation is governed by linear processes, but the crystal lattice spacing is determined by nonlinear terms. The wavelength mismatch can result in significant disorder behind the front that may or may not persist in the latter stage dynamics. We support these observations by presenting results from dynamical density functional theory calculations for simple one- and two-component two-dimensional systems of soft core particles.

Keywords Crystallization • Front motion • Dynamical density functional theory • Disordered systems • Glass formation • Soft matter

A.J. Archer (✉) • M.C. Walters
Department of Mathematical Sciences, Loughborough University, Loughborough, LE11 3TU, UK
e-mail: a.j.archer@lboro.ac.uk; m.walters@lboro.ac.uk

U. Thiele
Institut für Theoretische Physik, Westfälische Wilhelms-Universität Münster, Wilhelm Klemm Str. 9, D-48149, Münster, Germany

Center of Nonlinear Science (CeNoS), Westfälische Wilhelms Universität Münster, Corrensstr. 2, 48149, Münster, Germany
e-mail: u.thiele@uni-muenster.de

E. Knobloch
Department of Physics, University of California at Berkeley, Berkeley, CA 94720, USA
e-mail: knobloch@berkeley.edu

Y. Nishiura, M. Kotani (eds.), *Mathematical Challenges in a New Phase of Materials Science*, Springer Proceedings in Mathematics & Statistics 166,
DOI 10.1007/978-4-431-56104-0_1

1 Introduction

Solids with a well-ordered crystalline structure have numerous applications in materials science. In this article we focus on some of the considerations that determine whether a crystal grown from a supercooled liquid will be essentially defect-free or will contain substantial disorder that will modify its material properties. We are interested in particular in the consequence of quenching a uniform liquid to a temperature at which (a) the thermodynamically stable state is a crystalline solid, and (b) the supercooled liquid is linearly unstable with respect to growing density perturbations. This occurs when the temperature quench is sufficiently deep. When the temperature of the liquid is quenched only a little below the freezing temperature, the crystal forms instead via nucleation and growth [1–4]. However, if the temperature quench is sufficiently deep, then the uniform liquid is unstable and any small perturbations in the density grow spontaneously [5–7]. This is often referred to as the spinodal regime. One expects that in this case the perturbations will evolve into a well-ordered crystalline solid but whether this is the case depends crucially on the speed of the crystallisation front that develops from an initial small amplitude density perturbation [7, 8]. Of course we anticipate that the front speed will be slow for shallow quenches and faster for deep quenches, an expectation we confirm here. The front speed plays a crucial role since the wavenumber of the density modulations deposited behind the crystallisation front is in fact determined dynamically and so may differ from the wavenumber of the crystal in thermodynamic equilibrium. We examine here the physical mechanisms responsible for the speed of the crystallisation front and distinguish between the so-called pulled fronts whose speed is determined by linear processes and pushed fronts whose speed is determined nonlinearly [9].

The fact that the wavelength of the density modulations deposited by a pulled crystallisation front differs in general from the length scale of the equilibrium crystal is important in determining the nature of the solid structure that is formed. If the length scales are similar, then behind the front few defects are generated as the solid relaxes to equilibrium. In contrast, if the resulting wavelength mismatch is large, then the defect density and disorder in the solid are greatly enhanced. We find in general that this is particularly so for deep quenches where the front speed is high.

We elucidate these notions by studying the solidification of model two-dimensional (2D) systems of soft-core particles. We use a simple mean-field dynamical density function theory (DDFT) [10–14] that has been shown to be quite accurate in reproducing Brownian dynamics computer simulations to demonstrate that a mismatch between the wavelength of the modulations deposited behind a solidification front and the length scale of the equilibrium crystal is indeed likely when the liquid is deeply quenched. We also observe that this mismatch results in a disordered structure behind the front. In systems composed of only one species of particles, the disorder subsequently largely disappears as the particles are able to rearrange, healing the majority of the defects. However, in binary mixtures this disorder may be frozen-in.

In this paper we briefly review the pertinent results from our earlier work [7], and then focus on characterizing the disorder created by fast crystallisation fronts in several systems of interest and the factors influencing whether the resulting disorder persists over long time scales. We emphasise that wavenumber selection via a moving front is a linear process only when the temperature quench is deep. Shallow quenches to a temperature just below where the liquid and crystalline phases coexist do not generate crystallisation fronts unless the crystal phase is nucleated; the resulting front is necessarily a pushed front. Furthermore, this type of front persists into the linearly unstable state until the quench becomes so deep that the speed of the pulled front exceeds the speed of the pushed front [7].

Propagation of soft matter crystallisation fronts has been considered before, notably in Refs. [8, 15, 16] within the so-called phase-field crystal description in one spatial dimension and in Refs. [17–19] in two spatial dimensions, in both cases focusing on the properties of pulled fronts. The more accurate DDFT approach used here extends and generalises these results to both types of crystallisation fronts and to two spatial dimensions.

We should mention that the DDFT that we use assumes that the particles in the system follow overdamped but stochastic equations of motion – i.e. they are implicitly treated as colloidal particles immersed in a solvent that acts as a heat bath. Thus, after the quench the system is maintained at the quench temperature. This contact with a heat bath eliminates the effects of latent heat release. The absence of particle inertia implies that the dynamics are diffusive: no propagating excitations (phonons) are possible in such systems.

This paper is structured as follows: In Sects. 2 and 3 we describe the origin of the basic length scales involved in these processes. In Sects. 4 and 5 we summarise and illustrate the basic theory determining the speed of pulled fronts. In Sects. 6 and 7 we describe the DDFT theory for a single component fluid and the structures generated by crystallisation fronts moving with different speeds. Sections 8 and 9 compare and contrast these results with the results obtained for binary mixtures. The article concludes with a brief summary and a few concluding remarks.

2 Length Scales in Liquids and Solids

Consider a system of atoms, molecules, colloids, etc. (henceforth referred to as 'particles') which collectively exhibit a phase transition from a liquid to a crystalline state. Thermodynamic and structural properties of these two phases can in principle be found using classical density functional theory (DFT) [14, 20–22]. In DFT it is shown that there exists a functional $\Omega[\rho]$, together with a minimisation principle

$$\frac{\delta\Omega[\rho]}{\delta\rho} = 0. \tag{1}$$

The density profile $\rho^*(\mathbf{r})$ that solves this equation, i.e. that minimises $\Omega[\rho]$, is the density distribution of the system at equilibrium. Furthermore, $\Omega[\rho^*]$ is the thermodynamic grand potential of the system. Solving Eq. (1) for state points in the phase diagram where the liquid is the equilibrium phase yields a density profile that is uniform in space, $\rho(\mathbf{r}) = \rho_0$. In contrast, for the crystal phase, the density profile exhibits a regular array of peaks. From this density profile, quantities such as the crystal lattice spacing ℓ can be determined.

The functional $\Omega[\rho]$ is highly nonlinear, and quantities characterising the crystal, such as ℓ, depend on all terms in the functional. In contrast, quantities such as the static structure factor $S(k)$ [14] of the liquid only depend on the linear response of the liquid and so only depend on the terms in $\Omega[\rho]$ that are quadratic in the density fluctuation $\tilde{\rho} \equiv \rho - \rho_0$. Related to the static structure factor is the Fourier transform of the linear density response function $\chi(\mathbf{r} - \mathbf{r}')$, viz., $\hat{\chi}(k) = -(\rho_0/k_B T)S(k)$, that relates the change in the density $\delta\rho(\mathbf{r})$ to a change $\delta V_{ext}(\mathbf{r})$ in the external potential [20]:

$$\delta\rho(\mathbf{r}) = -\int d\mathbf{r}'\chi(\mathbf{r},\mathbf{r}')\delta V_{ext}(\mathbf{r}'). \tag{2}$$

This formula applies for both uniform and non-uniform fluids; in particular, in the case of a uniform fluid with density ρ_0 and $V_{ext} = 0$ perturbed by a small amplitude external potential $\delta V_{ext}(\mathbf{r})$, Eq. (2) determines the resulting change in the density profile: $\delta\rho \equiv \rho - \rho_0 = \tilde{\rho}$.[1]

The main point of the above comments is to emphasise that quantities pertaining to the crystal, such as ℓ, depend on all terms in $\Omega[\rho]$, but quantities such as $S(k)$ and the dispersion relation $\omega(k)$ [7, 8, 12], only depend on the quadratic terms in $\tilde{\rho}$. We emphasise this point because when a uniform liquid is deeply quenched, the length scales of the density modulations that initially grow after the quench are determined by $\omega(k)$ and so only depend on the quadratic terms in $\tilde{\rho}$. In particular, the principal peak in the dispersion relation $\omega(k)$ determines the wavenumber of the density fluctuation that grows the fastest. The fact that this wavenumber is determined by a quantity that only depends on the quadratic terms in $\tilde{\rho}$ shows that these fastest growing modes need not have the equilibrium wavenumber $2\pi/\ell$, i.e. they do not necessarily generate the correct density modulations for a perfect equilibrium crystal. From these considerations, one can infer that a deeply quenched liquid may well produce a disordered solid, because the length scale of the fastest growing modes is not in general equal to ℓ. This argument does not address whether, as solidification proceeds, the system can rearrange and subsequently anneal all the defects generated in the initial stages of the solidification process to produce a perfect crystal. Nonetheless, the observation that the initial dynamics after the

[1]The result in Eq. (2) also applies to non-uniform liquids, i.e. to liquids initially at equilibrium with a density profile $\rho_{old}(\mathbf{r})$ in an external potential $V_{old}(\mathbf{r})$, disturbed by an infinitesimal change to the external potential, $V_{old}(\mathbf{r}) \to V_{new}(\mathbf{r})$. The resulting change in the density profile $\delta\rho(\mathbf{r}) \equiv \rho_{new}(\mathbf{r}) - \rho_{old}(\mathbf{r})$ is then also given by Eq. (2), where $\delta V_{ext}(\mathbf{r}) \equiv V_{new}(\mathbf{r}) - V_{old}(\mathbf{r})$.

quench do not in general produce density fluctuations of the correct length scale is an important observation.

3 Dispersion Relation

The time evolution of the density distribution $\rho(\mathbf{r}, t)$ in a fluid system of particles is given by the continuity equation

$$\frac{\partial \rho}{\partial t} = -\nabla \cdot \mathbf{j}, \tag{3}$$

where $\mathbf{j}$ is the current. This equation is simply a statement of the fact that the number of particles in the system is a conserved quantity. To solve this equation of course requires an expression for the current $\mathbf{j} \equiv \rho \mathbf{u}$, where $\mathbf{u}$ is the local fluid velocity. In general, we only have formal expressions for this quantity, and to actually calculate the fluid dynamics, approximations are required. For example, for colloidal fluids, where the particles move with stochastic (Brownian) equations of motion, the following approximation can be rather good:

$$\mathbf{j} = -\Gamma \rho \nabla \frac{\delta \Omega[\rho]}{\delta \rho}, \tag{4}$$

where $\Gamma = D/k_B T$ is a mobility coefficient, D is the diffusion coefficient, k_B is Boltzmann's constant and T is the temperature. This approximation is the central result in DDFT [10–14].

Consider now a fluid with uniform density ρ_0 with a superposed small amplitude density perturbation $\tilde{\rho} \equiv \rho - \rho_0$ that is either artificially imposed, for example, via an external field or a consequence of random thermal fluctuations. Equation (3) shows that the perturbation evolves according to

$$\frac{\partial \tilde{\rho}}{\partial t} = \mathscr{L} \tilde{\rho} + \mathscr{O}(\tilde{\rho}^2), \tag{5}$$

where $\mathscr{L}$ is an operator that is obtained by linearising the full dynamical equation (3). For example, in the colloidal case, where Eq. (4) is applicable, the operator $\mathscr{L} = D\nabla^2 - D\rho_0 \nabla^2 c^{(2)} \otimes$, where $\otimes$ denotes a convolution, i.e. $c^{(2)} \otimes \tilde{\rho} \equiv \int d\mathbf{r}' c^{(2)}(\mathbf{r} - \mathbf{r}')\tilde{\rho}(\mathbf{r}')$, and $c^{(2)}(\mathbf{r})$ is the Ornstein-Zernike pair direct correlation function [12, 14]. Linearising Eq. (5) and decomposing $\tilde{\rho}$ into a sum of different Fourier modes,

$$\tilde{\rho}(\mathbf{r}, t) = \sum_{\mathbf{k}} \hat{\rho}_{\mathbf{k}} e^{i\mathbf{k}\cdot\mathbf{r} + \omega(k)t}, \qquad k \equiv |\mathbf{k}|, \tag{6}$$

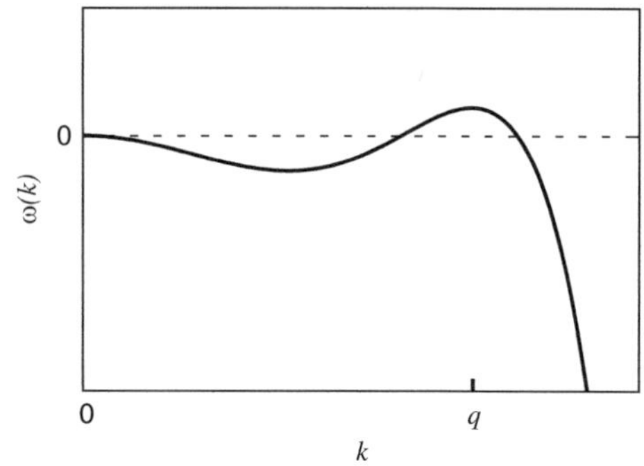

Fig. 1 Typical form of the dispersion relation $\omega(k)$ for a uniform liquid that is supercooled to the region of the phase diagram where it is linearly unstable against the growth of density modulations with wavenumber $k \approx q$

leads to a dispersion relation for the growth rate $\omega(k)$ of density fluctuations with wavenumber k. In the colloidal case described by Eq. (5) we obtain [7, 8, 12]:

$$\omega(k) = -Dk^2[1 - \rho_0 \hat{c}(k)], \tag{7}$$

where $\hat{c}(k)$ is the Fourier transform of $c^{(2)}(r)$. Note that for the equilibrium fluid, we also have the relation $S(k) = [1 - \rho_0 \hat{c}(k)]^{-1}$.

Equation (6) tells us that wavenumbers k for which $\omega(k) > 0$ grow while those for which $\omega(k) < 0$ decay. For a deeply supercooled liquid, it can be the case that $\omega(k)$ is positive for a band of wavenumbers around the value $k = q$, say. A sketch of the typical form of $\omega(k)$ for such a supercooled liquid is displayed in Fig. 1.

Thus in the early stages following a deep quench, we see the growth of density modulations with wavenumber $k \approx q$, leading to the appearance of the length scale $2\pi/q$ in the density distribution of the system. However, as emphasised above, this scale may differ from the equilibrium crystal lattice spacing ℓ.

4 Speed of Solidification Fronts: Marginal Stability Hypothesis

A further aspect that has not been mentioned in the discussion so far relates to the question of how do solidification fronts propagate into the unstable liquid and what length scale density modulations do such solidification fronts produce? In the case where the liquid is unstable the properties of the solidification front can often be determined from the marginal stability hypothesis [7, 8, 16, 23, 24]: Consider the leading edge of such a front, where the growing density modulations are still small in amplitude and suppose this front is advancing with velocity $\mathbf{v}$. In a reference frame that moves with the front, Eq. (5) becomes

$$\frac{\partial \tilde{\rho}}{\partial t} + \mathbf{v} \cdot \nabla \tilde{\rho} = \mathscr{L} \tilde{\rho} + \mathscr{O}(\tilde{\rho}^2) \tag{8}$$

and so the (complex) growth rate of density perturbations in this moving reference frame becomes $\omega_{\mathbf{v}}(k) = i\mathbf{k} \cdot \mathbf{v} + \omega(k)$. The marginal stability hypothesis posits that the density modulation at the leading edge of the front has zero growth rate in the comoving frame – if this mode were to have a positive growth rate, the front would be moving faster than the reference frame; if it were to have a negative growth rate the front would retract. In other words, the (complex) group velocity in the comoving reference frame must be zero and the growth rate of the density perturbations must also be zero, i.e.,

$$iv + \frac{d\omega(k)}{dk} = 0 \tag{9}$$

and

$$Re[i\mathbf{k} \cdot \mathbf{v} + \omega(k)] = 0, \tag{10}$$

where $k \equiv k_r + ik_i$ and $v = |\mathbf{v}|$. These equations can in fact be derived, under appropriate conditions, using the method of stationary phase applied to the longtime evolution of an infinitesimal spatially localized initial density perturbation [9]. Together they provide three conditions that are to be solved for the three unknowns, v, k_r and k_i. In the front region, the density profile $\tilde{\rho}(\mathbf{r}, t)$ takes the form $\rho_{\text{front}}(\boldsymbol{\xi}, t)$, where $\boldsymbol{\xi} \equiv \mathbf{r} - \mathbf{v}t$, since the moving front generates wavelengths in a periodic fashion. Without loss of generality we take the front to move along the x-axis, i.e. $\mathbf{v} = (v, 0, 0)$, implying that $\tilde{\rho}(\mathbf{r}, t) \sim \exp(-k_i x)\sin(k_r(x - vt) + Im[\omega(k)]t)$. Thus k_r determines the wavelength of the density modulations in the front. More importantly, if no phase slips take place, then the wavenumber of the density modulations left behind by the front is [7, 8, 16, 24]:

$$k^* = k_r + \frac{1}{v} Im[\omega(k)]. \tag{11}$$

Thus k^* is determined by the form of $\omega(k)$, which is obtained from the linearised equation (5). Therefore the wavelength $2\pi/k^*$ of the density modulation created behind the advancing front also differs in general from the equilibrium crystal lattice spacing ℓ, just like the wavelength associated with the fastest growing mode.

Thus even if the length scale determined by the maximum in $\omega(k)$ is the same as the equilibrium lattice spacing for the crystal, as is the case in the simple PFC theory for the crystal, solidification fronts advancing into a deeply supercooled liquid will still generate density modulations with a distinct wavelength, requiring substantial subsequent rearrangements of the system in order to form a defect-free crystal without strain.

5 Marginal Stability Calculation for a Simple Model

In this section we perform the marginal stability calculation to obtain the front speed v and wavenumber k^* of the density modulations created behind the front to show how these quantities depend on the dispersion relation. We approximate the dispersion relation by making a Taylor expansion around the wavenumber corresponding to the principal peak and truncating after the k^3 term,

$$\omega(k) \approx \omega_m - a(k-q)^2 - b(k-q)^3, \tag{12}$$

where $\omega_m = \omega(k = q)$ is the maximum growth rate. The coefficient $a > 0$ is related to the width of the principal peak while b measures its asymmetry around the peak wavenumber $k = q$.

Substituting Eq. (12) into Eq. (9) we obtain:

$$iv - 2a(d + ik_i) - 3b(d + ik_i)^2 = 0, \tag{13}$$

where we have written $d = k_r - q$. Separating the real and imaginary parts of this equation, we obtain the following expressions for the front speed and the imaginary part of k:

$$v = 2(a + 3bd)k_i \tag{14}$$

$$k_i = \sqrt{\frac{2ad}{3b} + d^2}. \tag{15}$$

Substituting Eq. (12) into Eq. (10) we obtain:

$$Re[iv(k_r + ik_i) + \omega_m - a(d + ik_i)^2 - b(d + ik_i)^3] = 0 \tag{16}$$

giving

$$-vk_i + \omega_m - a(d^2 - k_i^2) - b(d^3 - 3dk_i^2) = 0. \tag{17}$$

Inserting Eqs. (14) and (15) into Eq. (17) we obtain a cubic equation to be solved for d. However, for present illustrative purposes it is instructive to proceed analytically on the assumption that d is a small quantity. In this case Eqs. (14) and (15) become

$$v \approx 2ak_i \tag{18}$$

$$k_i \approx \sqrt{\frac{2ad}{3b}}. \tag{19}$$

Linearization of Eq. (17) in d now leads to $d = 3b\omega_m/2a^2$, i.e. to

$$k_r \approx q + \frac{3b\omega_m}{2a^2}. \tag{20}$$

This result shows that the wavenumber k_r of the density modulation in the advancing solidification front is not equal to the wavenumber of the fastest growing mode for the quenched uniform fluid, q, unless the peak of the dispersion relation is symmetric, i.e. unless $b = 0$. We also see that the difference between these two wavenumbers grows with ω_m, the magnitude of which is related to the degree of undercooling. The deeper the quench, the larger is ω_m. Moreover, inserting these results into Eq. (11) we obtain the wavenumber k^* of the modulations deposited behind the front:

$$k^* \approx q + \frac{b\omega_m}{2a^2}. \tag{21}$$

Thus the wavenumber k^* differs in general from the fastest growing wavenumber q, and neither of these wavenumbers is in general equal to $2\pi/\ell$ and so defects and disorder must be present shortly after a deep quench. Some systems are subsequently able to rearrange, but others are not, as we show in the subsequent sections for a particular model fluid composed of soft-core particles.

6 Model Fluid

We consider a 2D system of soft-core particles interacting via the so-called generalised exponential model of index n (GEM-n) pair-potential:

$$w(r) = \epsilon e^{-(r/R)^n}, \tag{22}$$

where the parameter $0 < \epsilon < \infty$ determines the energy penalty for a pair of particles to overlap completely, R is the radius of the particles and the exponent n determines the 'softness' of the potential. When $n = 2$, the potential varies slowly. In contrast, when n is large, as the separation distance r between a pair of particles is decreased, the potential increases rapidly from ≈ 0 to a value $\approx \epsilon$ over a short distance at $r = R$. Here, we consider the case when $n = 8$. Such soft potentials arise as the effective interaction potential between polymers or soft macromolecules in solution [25–37].

Our main reason for considering this model centres on the fact that the structure and phase behaviour (i.e. thermodynamics) of this model is well described by a rather simple approximation for the free energy. The grand potential of the system can be decomposed as follows [14, 20–22]:

$$\Omega[\rho(\mathbf{r})] = \mathscr{F}[\rho(\mathbf{r})] + \int d\mathbf{r}\rho(\mathbf{r})(V_{ext}(\mathbf{r}) - \mu), \tag{23}$$

where μ is the chemical potential and $\mathscr{F}[\rho(\mathbf{r})] = \mathscr{F}_{id}[\rho(\mathbf{r})] + \mathscr{F}_{ex}[\rho(\mathbf{r})]$ is the intrinsic Helmholtz free energy functional. The ideal gas contribution is

$$\mathscr{F}_{id}[\rho(\mathbf{r})] = k_B T \int d\mathbf{r}\rho(\mathbf{r}) \left(\ln[\rho(\mathbf{r})\Lambda^2] - 1\right), \tag{24}$$

where Λ is the thermal de Broglie wavelength and we use the following mean-field approximation for the excess contribution to the free energy [25]:

$$\mathscr{F}_{ex}[\rho(\mathbf{r})] = \frac{1}{2}\int d\mathbf{r} \int d\mathbf{r}' \rho(\mathbf{r}) w(|\mathbf{r} - \mathbf{r}'|)\rho(\mathbf{r}') \tag{25}$$

which has been widely used in studies of the structure and phase behaviour of soft-core systems [7, 25, 38–58].

The bulk phase diagrams for the 2D GEM-4 and GEM-8 systems were calculated in Ref. [7] (see also Ref. [59]) and are displayed in Fig. 2. The diagrams exhibit a liquid phase at low densities and/or high temperatures which freezes to form a novel cluster crystal phase as the temperature decreases or the density increases.

We assume that the particles move with overdamped stochastic (Brownian) equations of motion. Therefore, the time evolution of the non-equilibrium density profile $\rho(\mathbf{r}, t)$ may be determined using DDFT, i.e., using Eqs. (23), (24) and (25) together with Eqs. (3) and (4). Within the present mean-field approximation, the pair

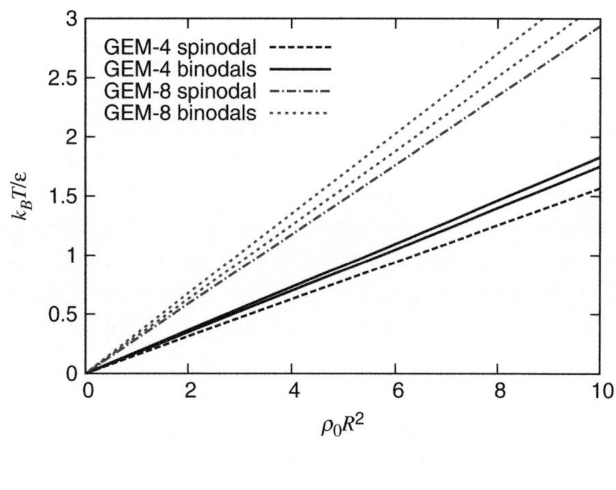

Fig. 2 The phase diagram of 2D GEM-4 and GEM-8 fluids. The binodals are lines of thermodynamic coexistence; the spinodals correspond to the onset of linear instability of the metastable uniform liquid

direct correlation function is [14, 20–22]:

$$c^{(2)}(|\mathbf{r}-\mathbf{r}'|) \equiv -\beta \frac{\delta^2 \mathscr{F}_{ex}[\rho(\mathbf{r})]}{\delta\rho(\mathbf{r})\delta\rho(\mathbf{r}')} = -\beta w(|\mathbf{r}-\mathbf{r}'|), \tag{26}$$

where $\beta = (k_B T)^{-1}$ and so the dispersion relation (7) has the following very simple form:

$$\omega(k) = -Dk^2[1 + \rho_0 \beta \hat{w}(k)], \tag{27}$$

where $\hat{w}(k)$ is the 2D Fourier transform of the pair potential $w(r)$. The threshold for linear instability of the uniform fluid is determined by $\omega(k = q) = 0$, where $q \neq 0$ is the wavevector at which $\omega(k)$ has a maximum, i.e., by $1 + \rho_0 \beta \hat{w}(q) = 0$. This leads to a very simple linear density dependence of the onset temperature: $k_B T = |\hat{w}(k = q)|\rho_0$, where the marginally stable wavenumber q at onset is determined by the condition

$$\left.\frac{d\hat{w}(k)}{dk}\right|_{k=q} = 0. \tag{28}$$

For the 2D GEM-8 system the onset wavenumber $q \approx 5.26/R$. Since $\hat{w}(k = q) \approx -0.294\epsilon R^2$ is independent of the density, the linear instability threshold is a straight line in the phase diagram:

$$\frac{k_B T}{\epsilon} \approx 0.294\rho_0 R^2 \tag{29}$$

(Fig. 2). In addition, the binodals along which the liquid and crystal phases coexist in thermodynamic equilibrium also appear to be straight lines in the phase diagram. This is not obvious a priori because the binodal calculation requires that one first obtains the crystal density profile, which is a highly nonlinear problem. However, fitting the numerically obtained binodals with a straight line proves to be an excellent approximation (Fig. 2). For example, for the GEM-8 fluid we find that the binodal for the crystal state at coexistence is given by $k_B T/\epsilon \approx 0.314\rho_0 R^2$ and that of the liquid is $k_B T/\epsilon \approx 0.339\rho_0 R^2$. Thus, when the temperature $k_B T/\epsilon = 1$, the density of the liquid at coexistence is $\rho_0 R^2 \approx 1/0.339 = 2.95$ while that of the crystal is $\rho_0 R^2 \approx 1/0.314 = 3.18$.

At thermodynamic coexistence, at the temperature $T = T_{\text{coex}}$ (and the chemical potential $\mu = \mu_{\text{coex}}$), a front between the crystal and the liquid state is stationary. However, on decreasing the temperature below $T = T_{\text{coex}}$ (or increasing μ above $\mu = \mu_{\text{coex}}$), the liquid state is no longer the equilibrium state. For a shallow quench, the liquid state remains linearly stable but a crystal can still grow if it is nucleated: only a crystal seed that is larger than the critical size grows and the interface (front) between the two phases advances at a well-defined speed v_{nl} determined by nonlinear processes [7, 60]. This (pushed) front propagation was studied in detail in Ref. [7] for the GEM-4 model. At $T = T_{\text{coex}}$, the front speed $v_{\text{nl}} = 0$; as the temperature T decreases below T_{coex} the speed v_{nl} increases with increasing quench depth $|T - T_{\text{coex}}|$.

If the quench is to a temperature $T < T_{\text{sp}}$, where T_{sp} is the temperature determined by Eq. (29) at which the uniform liquid becomes linearly unstable (i.e. the spinodal), then front propagation via linear processes is possible, with the speed v determined by the marginal stability analysis described in Sect. 4. However, as can be seen from Eqs. (18), (19) and (20), $v = 0$ at $T = T_{\text{sp}}$ since $\omega_m = 0$ at T_{sp}. As the temperature is decreased below T_{sp}, v increases but remains less than v_{nl} for small $|T - T_{\text{sp}}|$. In this regime the front remains a pushed front even though the liquid is already unstable [7, 60, 61]. However, the speed v increases faster than v_{nl} with decreasing T resulting in a crossover in speeds at temperature $T = T_{\text{x}}$, where $T_{\text{x}} < T_{\text{sp}} < T_{\text{coex}}$. At T_{x}, the two speeds are equal, $v = v_{\text{nl}}$, but for a sufficiently deep quench $v > v_{\text{nl}}$, and for these temperatures ($T < T_{\text{x}}$) it is the linear process that determines how the crystal state advances into the unstable liquid [7, 60, 61].

The variation of the speed of the crystallisation front with increasing chemical potential μ (at fixed temperature) is analogous to that described above for decreasing temperature (at fixed chemical potential). The metastable uniform liquid becomes linearly unstable at $\mu_{\text{sp}} > \mu_{\text{coex}}$ and for $\mu > \mu_{\text{sp}}$ front propagation via linear processes is possible. However, it is only when $\mu > \mu_{\text{x}} > \mu_{\text{sp}}$ that linear processes govern the propagation of the front and the front speed is determined by the marginal stability result. In Fig. 3a (see also Fig. 4 of Ref. [7]) we show for a GEM-4 fluid with temperature $k_BT/\epsilon = 1$ that for $\beta\mu > \beta\mu_{\text{x}} \approx 21$ the front speed obtained from solving the DDFT equations numerically in 2D does indeed agree precisely with the speed v predicted by the marginal stability analysis. Figure 3b (see also Fig. 6 of Ref. [7]) compares the wavenumber k^* deposited behind a front generated

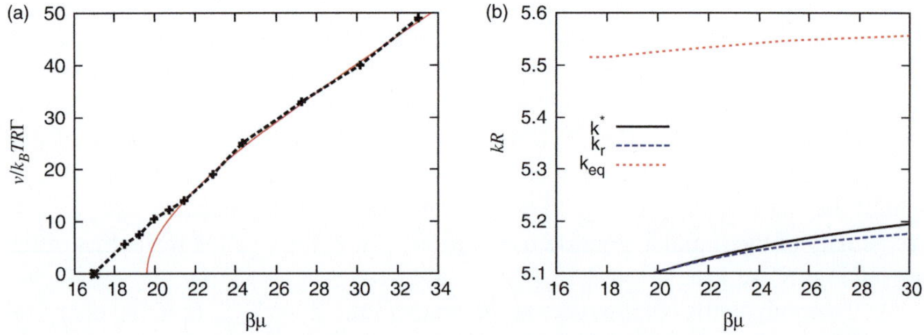

Fig. 3 (**a**) The front speed v as a function of the chemical potential μ for the GEM-4 fluid with temperature $k_BT/\epsilon = 1$. The *solid line* is the result of the marginal stability calculation while the symbols connected with a *dashed line* summarize the results from numerical simulations of the 2D DDFT equations. (**b**) The wavenumber k_r of the density modulation selected by the moving front, the wavenumber k^* deposited behind the front (both calculated from the marginal stability condition), and the wavenumber k_{eq} of the equilibrium crystal. The difference between k^* and k_{eq} implies that rearrangements behind the front are inevitable as the system seeks to minimise its free energy

by a deep quench ($\mu > \mu_x$, equivalently $T < T_x$) with the wavenumber k_{eq} corresponding to the equilibrium crystal lattice spacing. Our aim below is to explore the consequences of the dramatic difference between k^* and k_{eq} revealed in the figure for the subsequent evolution of the solid phase, and to demonstrate that it is responsible for the inevitable presence of defects and disorder.

7 Structures Formed After a Quench

In Fig. 4 we display a series of snapshots of the density profile calculated for a 2D GEM-8 fluid with bulk density $\rho_0 R^2 = 5$, quenched to the temperature $T^* = k_B T/\epsilon = 1$, where the uniform fluid is linearly unstable [cf. Eq. (29)].

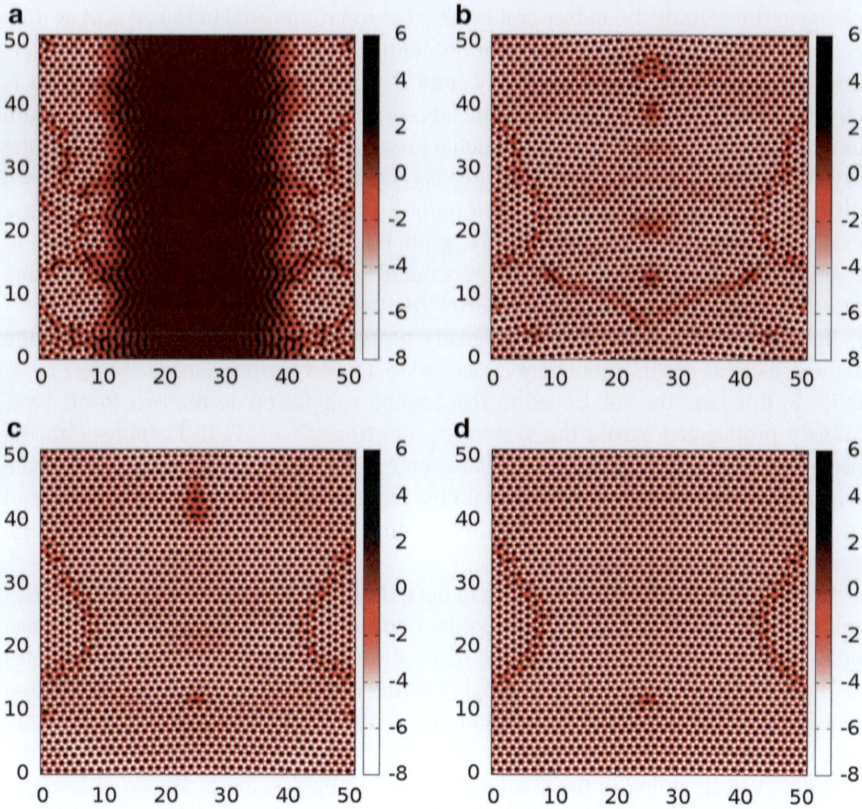

Fig. 4 The logarithm of the 2D density profile, $\ln(\rho(\mathbf{r}, t)R^2)$, at times $t^* = 1, 2, 10$ and 100 (going from top left to bottom right) obtained from DDFT for $\rho_0 R^2 = 5$ and temperature $k_B T/\epsilon = 1$. Plotting the logarithm of the density, rather than the density itself, allows one to see the grain boundaries and defects more clearly. The solidification front was initiated at time $t^* = 0$ along the line $x = 0$, or equivalently $x = 51.2R$

The times are given in Brownian time units, $t^* \equiv k_B T \Gamma R^2 t$, where $t^* \sim 1$ is the time it takes for a particle to diffuse a distance $\sim R$. Solidification is initiated by imposing at time $t = 0$ small amplitude random fluctuations along the line $x = 0$ on an otherwise uniform density profile with periodic boundary conditions. This produces a pair of solidification fronts propagating out to the left and to the right from this line. These fronts can be observed in the top left time $t^* = 1$ density profile in Fig. 4. Ahead of the front one observes stripe-like oscillations in the liquid, as predicted by the marginal stability analysis (see Refs. [7, 8] for other examples of similar fronts). Behind this stripe-like precursor the crystal grows but contains defects, a consequence of the mismatch between the dynamically generated length scale $2\pi/k^*$ and the equilibrium length scale ℓ: to lower its free energy and form the equilibrium crystal structure, the system must perform significant rearrangements. These rearrangements occur locally, generating defects behind the propagating crystallisation front. By time $t^* = 2$, the two fronts have collided owing to the periodic boundary conditions (see top right panel in Fig. 4) and as time proceeds the crystal density distribution continues to rearrange, removing defects in order to lower the free energy. By time $t^* = 100$ (bottom right) the system is largely occupied by a well-ordered hexagonal crystal with the correct equilibrium length scale ℓ. However, even after such a long time, there is still a region where the crystal orientation is not aligned with the crystal in the centre of the system – i.e. two different crystal grains remain present in the system, with defects remaining on the grain-boundary between these two. By 'grain boundaries' we refer to regions where regular hexagonal ordering is absent as occurs along the boundaries of neighbouring regions of misaligned hexagonal ordering (the 'grains').

In Fig. 5 we display a series of density profile snapshots for a system with the same density as in Fig. 4, but here quenched to a lower temperature, $T^* = k_B T/\epsilon = 0.75$. In this case the solidification front speed v is faster, so the two fronts have already propagated across the system by the time $t^* = 1$. In keeping with the prediction in Eq. (11), the mismatch between $2\pi/k^*$ and the equilibrium length scale ℓ increases with quench depth (i.e. with the parameter ω_m) and so more defects and disorder are produced in the system. Over time these mostly anneal out, although some persist for very long times.

To quantify the degree of ordering in the density profiles, we perform a Delauney triangulation [7, 8, 62, 63] on the points corresponding to all the maxima in the density profile. We then calculate the distribution function $p(\theta)$ of the bond angles θ in these triangles. Perfect hexagonal ordering results in a Delauney triangulation consisting entirely of equilateral triangles, corresponding to a single sharp peak at 60° in $p(\theta)$. In Fig. 6 we display $p(\theta)$ at times $t^* = 2$, 10 and 100 after the quench, corresponding to the results displayed in Figs. 4 and 5. We do indeed see a single peak in $p(\theta)$, centred at 60°. However, the peak also exhibits 'shoulders' on either side, particularly for short times after the quench, indicating the presence of strain and disorder at early times. These shoulders diminish over time, but never completely disappear. Comparing the results for the temperature $T^* = 1$ (left plot in Fig. 6) with those for $T^* = 0.75$ (right plot in Fig. 6), we observe that for shallower

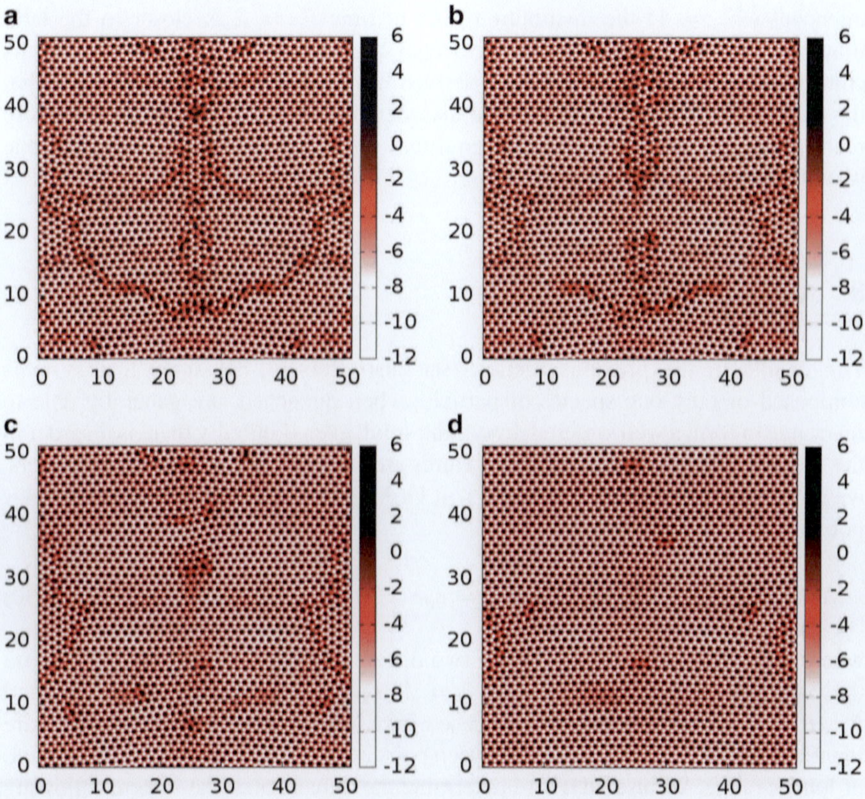

Fig. 5 Same as Fig. 4, but for the lower temperature $k_B T/\epsilon = 0.75$

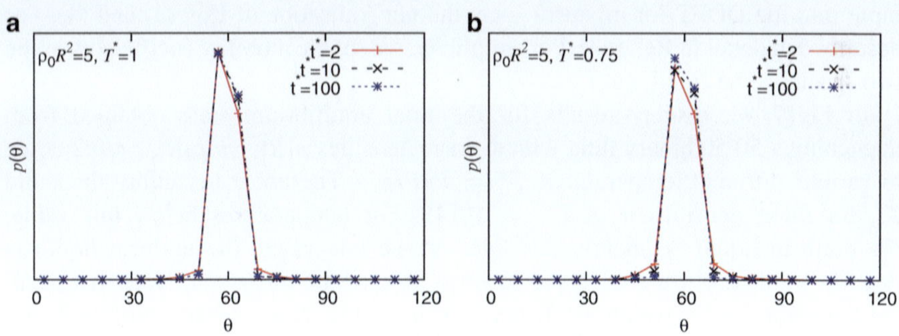

Fig. 6 The bond angle distribution $p(\theta)$, calculated from Delauney triangulation on the density profiles displayed in Figs. 4 and 5. The distribution exhibits a single peak at 60° due to the predominance of hexagonal ordering although shoulders on either side of this peak are present at early times due to the disorder in the system. These decrease over time, as the system rearranges to remove defects and lower the free energy

quenches ($T^* = 1$) the distribution $p(\theta)$ at time $t^* = 2$ is closer to the later time equilibrium distribution than for deeper quenches ($T^* = 0.75$). These results confirm and quantify the impression derived from the 2D density profiles, i.e. that the deeper the quench, the greater the disorder created in the system and the more rearrangements that are required to equilibrate the system. This is also what one would expect on the basis of Eq. (11).

8 A Binary Mixture

The results in the previous section (see also Refs. [7, 8]) show that systems composed of only one species of particle, when quenched, are generally able to rearrange to form a well-ordered crystalline solid, even if initially there is disorder in the system. However, since binary mixtures are far more likely to be glass formers, we now consider a binary mixture of soft GEM-8 particles, interacting via the pair potential [cf. Eq. (22)]:

$$w_{ij}(r) = \epsilon_{ij} e^{-(r/R_{ij})^8}, \tag{30}$$

where the indices $i, j = 1, 2$ label the two different species of particles. We choose the pair potential parameters $\epsilon_{12}/\epsilon_{11} = 1$, $\epsilon_{22}/\epsilon_{11} = 1.5$, $R_{12}/R_{11} = 1$ and $R_{22}/R_{11} = 1.3$, corresponding to the case where the potentials $w_{11}(r)$ and $w_{12}(r)$ are identical, but the potential $w_{22}(r)$ between pairs of species 2 particles is stronger and of longer range. We use DDFT to determine the dynamics of the system following a quench of the uniform liquid to temperatures where it is linearly unstable and a hexagonal crystal is the equilibrium state. We approximate the free energy of the binary mixture using a two-component generalisation of Eqs. (23), (24) and (25), input into the DDFT for mixtures – i.e. the generalisation of Eqs. (3) and (4) – as described in detail in Ref. [64]. For simplicity, we set the mobility coefficients of the two species to be equal, i.e. $\Gamma_1 = \Gamma_2 = \Gamma$.

In Fig. 7 we display results for the final equilibrium state obtained from quenching a 50:50 binary fluid with average densities $\bar{\rho}_1 R_{11}^2 = \bar{\rho}_2 R_{11}^2 = 0.2$ down to various different temperatures, $T^* = k_B T/\epsilon_{11}$. The linear instability threshold $T_{\rm sp}$ for these densities is at $T^* = 0.149$. For temperatures below this value, the uniform liquid is linearly unstable. Above this value, the uniform liquid is linearly stable and for the crystal to form, it must be nucleated. Note that at all the temperatures for which we display results in Fig. 7, the density profiles of the two species cease evolving after a time of $t^* \approx 100 - 200$. In Fig. 7 we plot the quantity $[\rho_1(\mathbf{r}) - \rho_2(\mathbf{r})]R_{11}^2$ and regions where $\rho_1 > \rho_2$ are coloured black while regions where $\rho_1 < \rho_2$ are coloured white. Regions coloured red are where either (i) both densities are small, or (ii) the two densities are equal in magnitude, but not necessarily small. It is worth observing that around most of the black density peaks there is a pale 'ring'. This is because wherever there is a peak in the density of

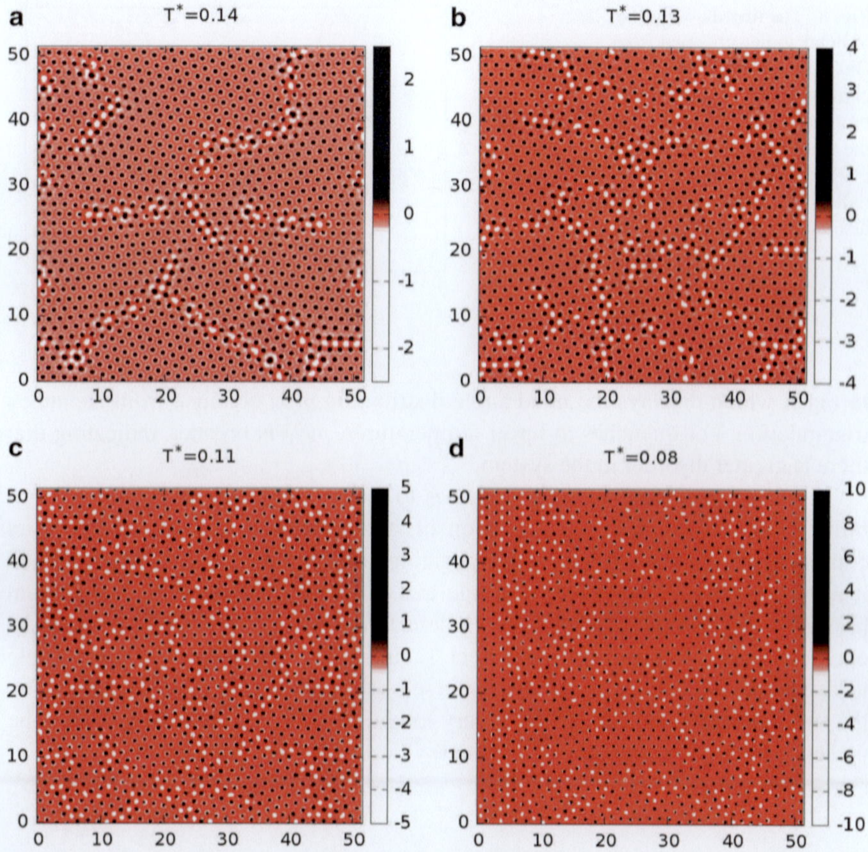

Fig. 7 The density difference $[\rho_1(\mathbf{r}) - \rho_2(\mathbf{r})]R_{11}^2$ after time $t^* = 100$ (when the profiles are almost stationary) obtained from DDFT with average densities $\bar{\rho}_1 R_{11}^2 = \bar{\rho}_2 R_{11}^2 = 0.2$ and the pair potential parameters $\epsilon_{12}/\epsilon_{11} = 1$, $\epsilon_{22}/\epsilon_{11} = 1.5$, $R_{12}/R_{11} = 1$ and $R_{22}/R_{11} = 1.3$. The dimensionless temperatures $T^* = k_B T/\epsilon_{11}$ are given above each figure. Plotting this quantity reveals the location of the species 2 particles (regions coloured *white*) which aggregate at the defects and grain boundaries. The grain size increases with the temperature T^*

species 1, there is also normally a peak in the density of species 2. The peaks in $\rho_2(\mathbf{r})$ are, however, broader than the peaks in $\rho_1(\mathbf{r})$ but lower in height, hence the 'ring'. They are lower in height because many of the species 2 particles gather at the grain boundaries; these are the white peaks in the profiles in Fig. 7. The eye can easily pick out the grain boundaries in these plots, from the lines of white peaks. We also observe that the density peaks at higher temperatures are broader than those at the lower temperatures (cf. Lindemann's criterion for melting [14]). Comparing the different profiles in Fig. 7 we see that the deeper the quench, the smaller the size of the crystalline grains and the more disorder is present, since there are more grain boundaries in the system. This observation is supported by the results

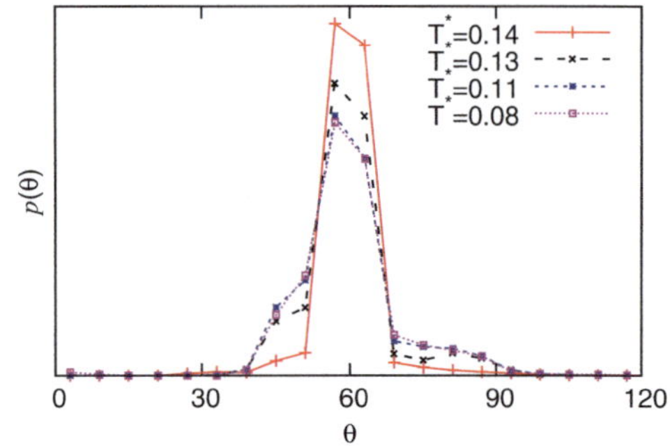

Fig. 8 The distribution $p(\theta)$ of bond angles obtained from Delauney triangulation, corresponding to the density profiles in Fig. 7. Note that as the temperature decreases $p(\theta)$ becomes broader, indicating that there is greater disorder in the system

in Fig. 8 which displays the bond angle distribution $p(\theta)$ obtained from Delauney triangulation. For quenches to lower temperatures, $p(\theta)$ is broader, indicating that there is greater disorder in the system.

These results show that deep quenches to temperatures below the linear instability threshold result in the formation of states that are highly disordered – a consequence of the fact that the propagating solidification front deposits behind it a dynamically generated length scale that differs from that of the equilibrium hexagonal crystal, thereby introducing defects into the system. In contrast to the one-component system discussed in Sect. 7, binary mixtures of particles are not able to rearrange as easily as in a monodisperse system and so the disorder created by the solidification front persists for a very long time (indeed indefinitely, within the present simple mean-field DFT treatment).

9 System with Competing Crystal Structures

We now present results for a different GEM-8 binary mixture, which has pair potential parameters $\epsilon_{11} = \epsilon_{12} = \epsilon_{22}$, $R_{12}/R_{11} = 1$ and $R_{22}/R_{11} = 1.5$. This particular system was first considered in Ref. [7], and we provide here additional detail. For fixed (sufficiently high) total density $\bar{\rho} \equiv \bar{\rho}_1 + \bar{\rho}_2$, where $\bar{\rho}_1$ and $\bar{\rho}_2$ are the average densities of species 1 and 2, respectively, the system exhibits several different equilibrium crystal structures depending on the concentration $\Phi \equiv \bar{\rho}_1/\bar{\rho}$ of species 1. Examples of these are displayed in Fig. 9. All the profiles displayed correspond to local minima of the free energy, but we have not checked whether they are the global minima at the given state points. At low values of Φ we observe a hexagonal crystal, such as that displayed in Fig. 9a. However, for larger values of Φ, the system can form a binary square crystal structure, such as that displayed in Fig. 9c. Alternatively, it can form a binary hexagonal lattice structure, such as that displayed in Fig. 9b,d. For high values of Φ the system forms a different simple hexagonal lattice, an example of which is displayed in Fig. 9e. In this hexagonal crystal the minority species particles occupy the same lattice sites as the majority

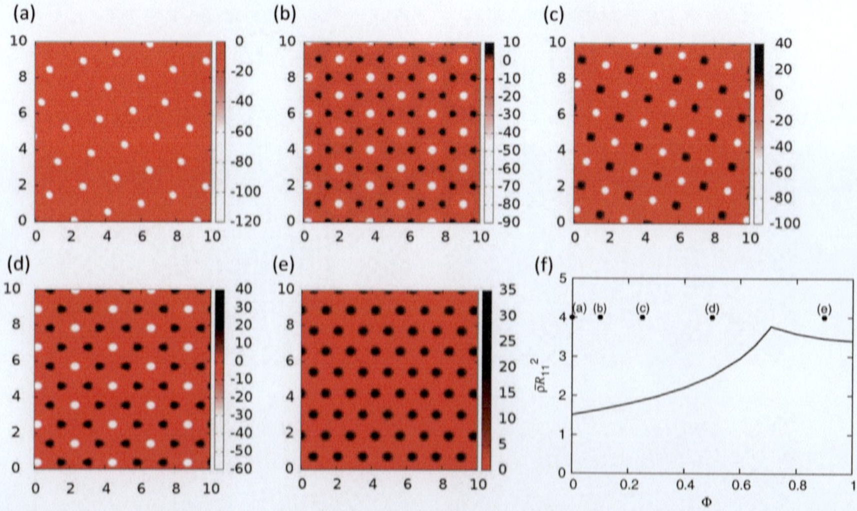

Fig. 9 Equilibrium crystal structures for the GEM-8 mixture with $\beta\epsilon_{ij} = \beta\epsilon = 1$ for all $i, j = 1, 2$, $R_{22}/R_{11} = 1.5$, $R_{12}/R_{11} = 1$ for concentrations **(a)** $\Phi = 0$, **(b)** 0.1, **(c)** 0.25, **(d)** 0.5, **(e)** 0.9 and average total density $\bar{\rho}R_{11}^2 = 4$. We plot the quantity $[\rho_1(\mathbf{r}) - \rho_2(\mathbf{r})]R_{11}^2$, so regions where $\rho_1(\mathbf{r}) > \rho_2(\mathbf{r})$ are coloured *black* while those where $\rho_1(\mathbf{r}) < \rho_2(\mathbf{r})$ are coloured *white*. All profiles correspond to local minima of the free energy, but we have not checked whether they correspond to global minima at the given state points. In **(f)** we display the linear stability threshold density as a function of concentration Φ, for fixed temperature $k_BT/\epsilon = 1$

species particles, in contrast to the lattice structures in (b)–(d). See Ref. [7] for further details.

We use DDFT to determine the structures that are formed when the uniform liquid mixture is quenched. A solidification front is initiated along the line $x/R_{11} = 25$ at time $t = 0$ by adding a small random value along this line. The fact that there are several competing crystal structures, in conjunction with the fact that we quench to state points that are far from the linear instability threshold (i.e. we perform a deep quench) means that the structures that are formed are highly disordered.

In Fig. 10 we present results from quenches for three systems with total density $\bar{\rho}R_{11}^2 = 8$ but different concentrations: $\Phi = 0.25$ (top), $\Phi = 0.5$ (middle) and $\Phi = 0.75$ (bottom).[2] Instead of displaying the density profiles calculated using DDFT, we plot the locations of all maxima in the total density profile $\rho(\mathbf{r}) \equiv \rho_1(\mathbf{r}) + \rho_2(\mathbf{r})$ exceeding $50R_{11}^{-2}$ at the peak. In addition, we colour-code these points, according to the local structure. The criteria for this are obtained via a Delauney triangulation on the points. Points where the local ordering is square are coloured black, points where there is hexagonal order are coloured red and the density peaks with neither local

[2]The early time profile displayed in the top right panel in Fig. 12 of Ref. [7] shows the corresponding plot for $\Phi = 0.25$ instead of $\Phi = 0.5$ as labelled there. This error is corrected here in Fig. 10.

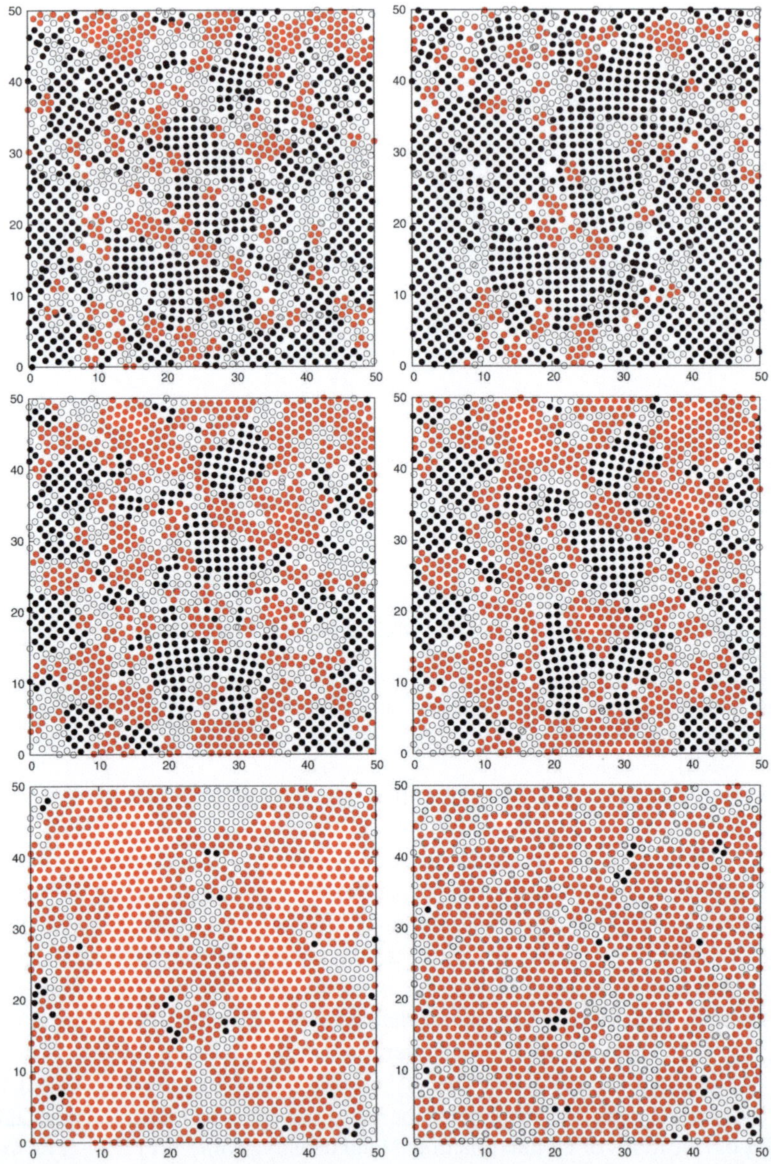

Fig. 10 The peaks in the density profile formed by a solidification front initiated along the line $x = 25$ at time $t = 0$. The system is a GEM-8 mixture with $\beta\epsilon_{ij} = 1$ for all $i, j = 1, 2$, $R_{22}/R_{11} = 1.5$ and $R_{12}/R_{11} = 1$ and average total density $(\bar{\rho}_1 + \bar{\rho}_2)R_{11}^2 \equiv \bar{\rho}R_{11}^2 = 8$. The results in the top row are for $\Phi = 0.25$, in the middle row for $\Phi = 0.5$ and the bottom row for $\Phi = 0.75$. In each case the plot on the left is for an early time $t^* = 1.6$, shortly after the solidification front has exited the domain and before the structure has had time to relax, while that on the right is for a later time (*top right* and *bottom right* are for $t^* = 40$, whilst the *middle right* is for $t^* = 400$). The density maxima are colour-coded according what kind of triangle they belong to in a Delauney triangulation: right-angled are *black*, equilateral are *red* and scalene are *open circles*. Portions of the hexagonal crystal is *red*, whilst the competing square crystal structure is *black*

ordering are shown as open circles.[3] In Fig. 10 we display on the left the structure formed at time $t^* = 1.6$, i.e., shortly after the solidification front has exited the box, and also the structure at a much later time (at time $t^* = 40$ for $\Phi = 0.25$ and $\Phi = 0.75$ and the time $t^* = 400$ for the case $\Phi = 0.5$ displayed on the middle right in Fig. 10). The structures that are formed are all highly disordered.

In the $\Phi = 0.25$ case displayed along the top row in Fig. 10 the structure that forms initially is a mixture of square and hexagonal ordering, but with a predominance of squares. There are also regions containing neither structure, largely on the grain boundaries. Over time, the regions of square ordering grow at the expense of the regions with hexagonal ordering. The density profiles continue to evolve very slightly after $t^* = 40$, but the displayed structure is very similar in its statistical properties to the final equilibrium structure.

The bond angle distribution $p(\theta)$ corresponding to the $\Phi = 0.25$ profiles is displayed in the top panel of Fig. 11. This distribution has two main peaks at 45° and 90°, reflecting significant square ordering in the system. Such squares yield right-angle triangles in the Delauney triangulation. Initially, at time $t^* = 1.6$, these peaks are rather broad, but over time they become sharper, reflecting the presence of growing domains of a well-ordered square crystal. The hexagonal ordering in the system is reflected by the fact that the peak in $p(\theta)$ at 45° has a 'shoulder' on it, extending out to 60°. This shoulder diminishes in height over time, but does not completely disappear, reflecting the fact that small regions of hexagonal ordering persist.

In the middle row of Fig. 10, we display the density peaks formed after quenching a uniform liquid with concentration $\Phi = 0.5$. Once again the structure that is formed contains regions of both square and hexagonal ordering. The size of both of these types of domains grows over time. In the middle right of Fig. 10 we display the density peaks at time $t^* = 400$, after which the density profiles cease to evolve. The corresponding bond angle distribution $p(\theta)$ is displayed in the middle panel of Fig. 11. This distribution has three peaks at 45°, 60° and 90°, reflecting the presence of a mixture of squares and hexagons. These peaks are rather broad, reflecting the significant disorder in the system.

In the bottom row of Fig. 10, we display the structure formed from quenching a concentration $\Phi = 0.75$ uniform liquid. We see that hexagonal ordering dominates, a fact confirmed by the large main peak in $p(\theta)$ located at 60° displayed in the lower panel of Fig. 11. What is particularly remarkable about these $\Phi = 0.75$ results is that the peak at 60° is actually sharper at the early times ($t^* = 1.6$) than later times ($t^* = 40$): over time the peak broadens! This is due to the fact that in this case

[3]The specific criterion for deciding to which subset a given density peak belongs is as follows: After performing the Delauney triangulation on the set of peaks, we consider each triangle. The corner angles are θ_1, θ_2 and θ_3. The triangle is defined as equilateral if $|\theta_i - \theta_j| < 5°$ for all pairs $i, j = 1, 2, 3$. The vertices of these triangles are coloured black. Triangles are defined as right-angled if for the largest angle θ_1 we have $|\theta_1 - 90°| < 5°$ and for the other two angles $|\theta_2 - \theta_3| < 5°$. The vertices of these triangles are coloured red. All remaining vertices which fall into neither of these categories are displayed as open circles.

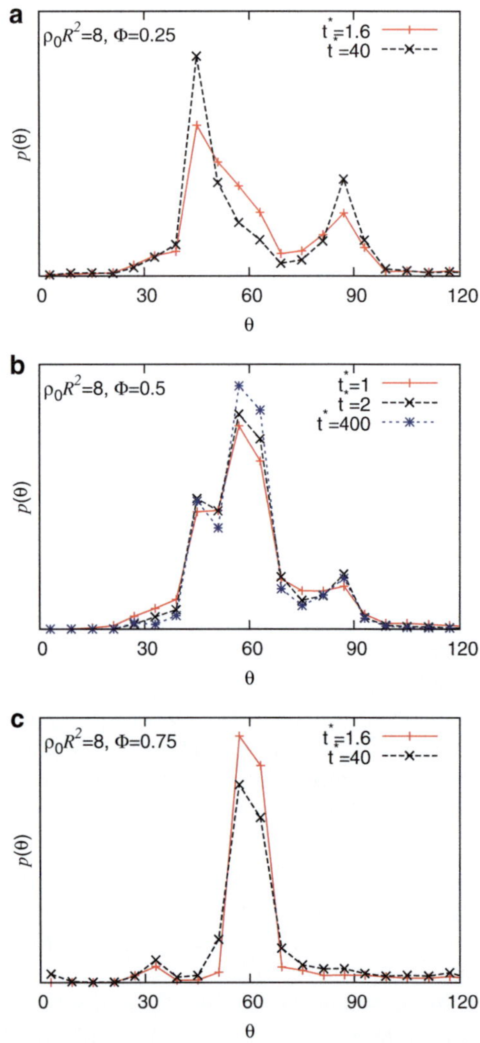

Fig. 11 The distribution $p(\theta)$ of bond angles obtained from Delauney triangulation, corresponding to the results in Fig. 10, which are for $\bar{\rho}R_{11}^2 = 8$ and concentration $\Phi = 0.25$ (*top*), $\Phi = 0.5$ (*middle*) and $\Phi = 0.75$ (*bottom*). In the *top plot*, the peaks in the distribution at 45° and 90° show that the system is largely composed of square ordered local structure. In contrast, the *bottom plot*, for $\Phi = 0.75$, has just one main peak in the distribution at 60° showing that the system is largely composed of hexagonal ordered local structure. In the *middle plot*, which is for $\Phi = 0.5$, $p(\theta)$ has peaks in the distribution at 45°, 60° and 90° showing that the system contains both square ordered local structure and also hexagonal local ordering, as one can also see from Fig. 10

the solidification front produces modulations with wavenumber k^* that is close to the wavenumber for the hexagonal crystal structure. However, these do not match exactly so that the hexagonal crystal that is initially formed is strained. Over time, the system lowers the free energy by *introducing* defects which alleviate the strain. These defects lead in turn to the broadening of the peak in $p(\theta)$ at 60°.

10 Concluding Remarks

In this paper we have discussed a mechanism that results in the formation of disordered structures, when a liquid is deeply quenched to temperatures where the thermodynamic equilibrium state is a well-ordered crystal. This occurs because solidification fronts in deeply quenched liquids propagate via a mechanism that generates periodic density modulations in the system with wavelength that is not necessarily the same as the wavelength required for an equilibrium crystal. The wavelength mismatch means that the formation of a well-ordered equilibrium crystal state requires significant rearrangements after the front has passed. In monodisperse one-component systems, these rearrangements should generally be possible; this is certainly the case in the model fluid studied here. However, for polydisperse systems or multi-component mixtures, such as the binary mixtures studied here, these rearrangements are frustrated and in some cases hindered by the fact that there is a variety of particle sizes in the system.

We should emphasise that the front propagation mechanism focused on in this paper operates only when the quench is sufficiently deep, to temperatures below the crossover temperature T_x. Only for $T < T_x$ do solidification fronts propagate via the linear mechanism, with the speed v and wavenumber k^* determined by linear considerations. Above T_x the front speed is determined by nonlinear considerations and in this regime the structure formed behind the front is generally much better ordered.

The results presented here are for rather simple 2D model systems composed of soft particles. Nevertheless, we believe that the mechanism that we describe should be rather general, although much further work is required to determine the nature of crystallisation fronts in other deeply quenched systems, and in particular to determine whether one can reach the regime where the fronts propagate via the linear mechanism that we describe. In other systems, it may be the case that the speed v never overtakes v_{nl}. For example, it is not known whether this regime is physically accessible for particles with a hard core; it may be that this regime only arises at densities near or even beyond random close-packing.

As a final point, we should mention that the mechanism described here is not the only means of introducing disorder as liquids solidify. Other mechanism include: (i) Defects created by impurities. (ii) Different grains, with defects on the grain boundaries, generated in the nucleation regime when growing crystals with different orientation nucleated at different points in the system collide. (iii) Defects introduced by crystal growth under the influence of external forces or shear. (iv) Disordered materials produced in shallow quenches where the growing crystal forms dendritic type structures (diffusion limited growth), leading to the formation of crystal grains and defects.

Acknowledgements A.J.A. and U.T. thank the Center of Nonlinear Science (CeNoS) of the University of Münster for recent support of their collaboration. M.C.W. is supported by an EPSRC studentship. The work of E.K. was supported in part by the National Science Foundation under Grant No. DMS-1211953.

References

1. Debenedetti, P.G.: Metastable Liquids: Concepts and Principles. University Press, Princeton (1996)
2. Oxtoby, D.W.: Homogeneous nucleation: theory and experiment. J. Phys.: Condens. Matter **4**, 7627 (1992)
3. Oxtoby, D.W.: Nucleation of first-order phase transitions. Acc. Chem. Res. **31**, 91 (1998)
4. Sear, R.P.: Nucleation: theory and applications to protein solutions and colloidal suspensions. J. Phys.: Condens. Matter **19**, 033101 (2007)
5. Trudu, F., Donadio, D., Parrinello, M.: Freezing of a Lennard-Jones fluid: from nucleation to spinodal regime. Phys. Rev. Lett. **97**, 105701 (2006)
6. Klein, W., Leyvraz, F.: Crystalline nucleation in deeply quenched liquids. Phys. Rev. Lett. **57**, 2845 (1986)
7. Archer, A.J., Walters, M.C., Thiele, U., Knobloch, E.: Solidification in soft-core fluids: disordered solids from fast solidification fronts. Phys. Rev. E **90**, 042404 (2014)
8. Archer, A.J., Robbins, M.J., Thiele, U., Knobloch, E.: Solidification fronts in supercooled liquids: How rapid fronts can lead to disordered glassy solids. Phys. Rev. E **86**, 031603 (2012)
9. van Saarloos, W.: Front propagation into unstable states. Phys. Rep. **386**, 29 (2003)
10. Marconi, U.M.B., Tarazona, P.: Dynamic density functional theory of fluids. J. Chem. Phys. **110**, 8032 (1999)
11. Marconi, U.M.B., Tarazona, P.: Dynamic density functional theory of fluids. J. Phys.: Condens. Matter **12**, A413 (2000)
12. Archer, A.J., Evans, R.: Dynamical density functional theory and its application to spinodal decomposition. J. Chem. Phys. **121**, 4246 (2004)
13. Archer, A.J., Rauscher, M.: Dynamical density functional theory for interacting Brownian particles: stochastic or deterministic? J. Phys. A: Math. Gen. **37**, 9325 (2004)
14. Hansen, J.P., McDonald, I.R.: Theory of Simple Liquids, 4th edn. Academic, London (2013)
15. Liu, F., Goldenfeld, N.: Dynamics of phase-separation in block copolymer melts. Phys. Rev. A **39**, 4805 (1989)
16. Galenko, P.K., Elder, K.R.: Marginal stability analysis of the phase field crystal model in one spatial dimension. Phys. Rev. B **83**, 064113 (2011)
17. Chen, L.Y., Goldenfeld, N., Oono, Y., Paquette, G.: Selection, stability and renormalization. Physica A **204**, 113 (1994)
18. van Teeffelen, S., Backofen, R., Voigt, A., Löwen, H.: Derivation of the phase-field-crystal model for colloidal solidification. Phys. Rev. E **79**, 051404 (2009)
19. Emmerich, H., Löwen, H., Wittkowski, R., Gruhn, T., Tóth, G.I., Tegze, G., Gránásy, L.: Phase-field-crystal models for condensed matter dynamics on atomic length and diffusive time scales: an overview. Adv. Phys. **61**, 665 (2012)
20. Evans, R.: Nature of the liquid-vapor interface and other topics in the statistical-mechanics of nonuniform, classical fluids. Adv. Phys. **28**, 143 (1979)
21. Evans, R.: Fundamentals of Inhomogeneous Fluids. Dekker, New York (1992)
22. Lutsko, J.F.: Recent developments in classical density functional theory. Adv. Chem. Phys. **144**, 1 (2010)
23. Dee, G., Langer, J.S.: Propagating pattern selection. Phys. Rev. Lett. **50**, 383 (1983)
24. Ben-Jacob, E., Brand, H., Dee, G., Kramer, L., Langer, J.S.: Pattern propagation in nonlinear dissipative systems. Physica D **14**, 348 (1985)
25. Likos, C.N.: Effective interactions in soft condensed matter physics. Phys. Rep. **348**, 267 (2001)
26. Dautenhahn, J., Hall, C.K.: Monte Carlo simulation of off-lattice polymer chains: effective pair potentials in dilute solution. Macromolecules **27**, 5399 (1994)
27. Likos, C.N., Löwen, H., Watzlawek, M., Abbas, B., Jucknischke, O., Allgaier, J., Richter, D.: Star polymers viewed as ultrasoft colloidal particles. Phys. Rev. Lett. **80**, 4450 (1998)

28. Louis, A.A., Bolhuis, P.G., Hansen, J.P., Meijer, E.J.: Can polymer coils be modeled as "soft colloids"? Phys. Rev. Lett. **85**, 2522 (2000)
29. Bolhuis, P.G., Louis, A.A., Hansen, J.P., Meijer, E.J.: Accurate effective pair potentials for polymer solutions. J. Chem. Phys. **114**, 4296 (2001)
30. Jusufi, A., Dzubiella, J., Likos, C.N., von Ferber, C., Löwen, H.: Effective interactions between star polymers and colloidal particles. J. Phys.: Condens. Matter **13**, 6177 (2001)
31. Dzubiella, J., Jusufi, A., Likos, C.N., von Ferber, C., Löwen, H., Stellbrink, J., Allgaier, J., Richter, D., Schofield, A.B., Smith, P.A., Poon, W.C.K., Pusey, P.N.: Phase separation in star-polymer-colloid mixtures. Phys. Rev. E **64**, 010401(R) (2001)
32. Louis, A.A., Bolhuis, P.G., Finken, R., Krakoviack, V., Meijer, E.J., Hansen, J.P.: Coarse-graining polymers as soft colloids. Physica A **306**, 251 (2002)
33. Likos, C.N., Harreis, H.M.: Star polymers: from conformations to interactions to phase diagrams. Condens. Matter Phys. **5**, 173 (2002)
34. Götze, I.O., Harreis, H.M., Likos, C.N.: Tunable effective interactions between dendritic macromolecules. J. Chem. Phys. **120**, 7761 (2004)
35. Mladek, B.M., Fernaud, M.J., Kahl, G., Neumann, M.: On the thermodynamic properties of the generalized Gaussian core model. Condens. Matter Phys. **8**, 135 (2005)
36. Likos, C.N.: Soft matter with soft particles. Soft Matter **2**, 478 (2006)
37. Lenz, D.A., Blaak, R., Likos, C.N., Mladek, B.M.: Microscopically resolved simulations prove the existence of soft cluster crystals. Phys. Rev. Lett. **109**, 228301 (2012)
38. Archer, A.J., Evans, R.: Binary Gaussian core model: fluid-fluid phase separation and interfacial properties. Phys. Rev. E **64**, 041501 (2001)
39. Archer, A.J., Evans, R.: J. Phys.: Wetting in the binary Gaussian core model. Condens. Matter **14**, 1131 (2002)
40. Archer, A.J., Evans, R., Roth, R.: Microscopic theory of solvent-mediated long-range forces: influence of wetting. Europhys. Lett. **59**, 526 (2002)
41. Archer, A.J., Likos, C.N., Evans, R.: Binary star-polymer solutions: bulk and interfacial properties. J. Phys.: Condens. Matter **14**, 12031 (2002)
42. Archer, A.J., Likos, C.N., Evans, R.: J. Phys.: Soft-core binary fluid exhibiting a lambda-line and freezing to a highly delocalized crystal. Condens. Matter **16**, L297 (2004)
43. Götze, I.O., Archer, A.J., Likos, C.N.: Structure, phase behavior, and inhomogeneous fluid properties of binary dendrimer mixtures. J. Chem. Phys. **124**, 084901 (2006)
44. Mladek, B.M., Gottwald, D., Kahl, G., Neumann, M., Likos, C.N.: Formation of polymorphic cluster phases for a class of models of purely repulsive soft spheres. Phys. Rev. Lett. **96**, 045701 (2006)
45. Mladek, B.M., Gottwald, D., Kahl, G., Neumann, M., Likos, C.N.: Clustering in the absence of attractions: density functional theory and computer simulations. J. Phys. Chem. B **111**, 12799 (2007)
46. Moreno, A.J., Likos, C.N.: Diffusion and relaxation dynamics in cluster crystals. Phys. Rev. Lett. **99**, 107801 (2007)
47. Likos, C.N., Mladek, B.M., Gottwald, D., Kahl, G.: Why do ultrasoft repulsive particles cluster and crystallize? Analytical results from density-functional theory. J. Chem. Phys. **126**, 224502 (2007)
48. Mladek, B.M., Charbonneau, P., Likos, C.N., Frenkel, D., Kahl, G.: Multiple occupancy crystals formed by purely repulsive soft particles. J. Phys.: Condens. Matter **20**, 494245 (2008)
49. Likos, C.N., Mladek, B.M., Moreno, A.J., Gottwald, D., Kahl, G.: Cluster-forming systems of ultrasoft repulsive particles: statics and dynamics. Comput. Phys. Commun. **179**, 71 (2008)
50. Overduin, S.D., Likos, C.N.: Clustering in nondemixing mixtures of repulsive particles. J. Chem. Phys. **131**, 034902 (2009)
51. Overduin, S.D., Likos, C.N.: Phase behaviour in binary mixtures of ultrasoft repulsive particles. Europhys. Lett. **85**, 26003 (2009)
52. van Teeffelen, S., Moreno, A.J., Likos, C.N.: Cluster crystals in confinement. Soft Matter **5**, 1024 (2009)

53. Camargo, M., Moreno, A.J., Likos, C.N.: Dynamics in binary cluster crystals. J. Stat. Mech. Theor. Exp. **2010**, P10015 (2010)
54. Camargo, M., Likos, C.N.: Interfacial and wetting behaviour of phase-separating ultrasoft mixtures. Mol. Phys. **109**, 1121 (2011)
55. Nikoubashman, A., Kahl, G., Likos, C.N.: Flow quantization and nonequilibrium nucleation of soft crystals. Soft Matter **8**, 4121 (2012)
56. Carta, M., Pini, D., Parola, A., Reatto, L.: A density-functional theory study of microphase formation in binary Gaussian mixtures. J. Phys.: Condens. Matter **24**, 284106 (2012)
57. Pini, D.: A density-functional theory investigation of cluster formation in an effective-potential model of dendrimers. Trans. R. Norw. Soc. Sci. Lett. **3**, 99 (2014)
58. Archer, A.J., Rucklidge, A.M., Knobloch, E.: Quasicrystalline order and a crystal-liquid state in a soft-core fluid. Phys. Rev. Lett. **111**, 165501 (2013)
59. Prestipino, S., Saija, F.: Hexatic phase and cluster crystals of two-dimensional GEM4 spheres. J. Chem. Phys. **141**, 184502 (2014)
60. Hari, A., Nepomnyashchy, A.A.: Nonpotential effects in dynamics of fronts between convection patterns. Phys. Rev. E **61**, 4835 (2000)
61. Couairon, A., Chomaz, J.M.: Pattern selection in the presence of a cross flow. Physica D **108**, 236 (1997)
62. de Berg, M., Cheong, O., van Kreveld, M., Overmars, M.: Computational Geometry: Algorithms and Applications, 3rd edn. Springer-Verlag, Heidelberg (2008)
63. Robbins, M.J., Archer, A.J., Thiele, U., Knobloch, E.: Modelling fluids and crystals using a two-component modified phase field crystal model. Phys. Rev. E **85**, 061408 (2012)
64. Archer, A.J.: Dynamical density functional theory: binary phase-separating colloidal fluid in a cavity. J. Phys.: Condens. Matter **17**, 1405 (2005)

Topological Analysis of the Diblock Copolymer Equation

Thomas Wanner

Abstract We demonstrate how topological methods can be used to study pattern formation and pattern evolution in phase-field models of materials science. In the context of the diblock copolymer model for microphase separation, we will present new quantitative results on the microstructure topology during the initial phase separation from a homogeneous state, both for a deterministic and a stochastic version of the model. We also describe the long-term dynamics of the model and associated questions of multistability, which can be addressed using rigorous topological methods aimed at determining the structure of the global attractor of the system.

Keywords Diblock copolymers • Phase field model • Pattern formation • Homology • Nonlinear dynamics • Numerical continuation • Bifurcation • Computer-assisted proof

1 Introduction

In recent years, computational topology has increasingly been used as a tool for understanding high-dimensional, complex data generated both from experiment and from numerical simulations. Many of these studies involve point-cloud data sets or networks, and they have demonstrated that topological methods can provide significant new insight into applied problems. Also in the context of evolution equations, computational topology has been used successfully, both in the context of spatio-temporal chaos and for model validation [21, 22, 30]. It is therefore not surprising that these novel techniques have found their way into the materials science context as well. Mostly, they have been used for questions of pattern classification, where it has already been demonstrated that topological metrics can provide important insight into material studies [22, 47, 56]. More recently, topological methods have also been able to provide rigorous mathematical insight

T. Wanner (✉)
Department of Mathematical Sciences, George Mason University, Fairfax, VA 22030, USA
e-mail: twanner@gmu.edu

Y. Nishiura, M. Kotani (eds.), *Mathematical Challenges in a New Phase of Materials Science*, Springer Proceedings in Mathematics & Statistics 166,
DOI 10.1007/978-4-431-56104-0_2

into dynamical questions in phase-field models [31, 32, 55], or could make studies rigorous that started out purely numerical in nature [14, 27].

In the current paper, we try to survey some of the methods described above with the hope of making them accessible to a wider audience. To focus our discussion, we concentrate throughout on the following specific phase-field model. One of the basic models for microphase separation in diblock copolymers has been described in [42], and in its original form was proposed by Ohta and Kawasaki [43] and Bahiana and Oono [1], see also [7, 8]. If one considers a material constrained to the bounded domain $\Omega \subset \mathbb{R}^d$, then their model considers a free-energy functional for the relative macroscopic monomer concentration u, i.e., the difference between the two monomer volume fractions, given by

$$E_{\epsilon,\sigma}[u] = \int_\Omega \left(\frac{\epsilon^2}{2} |\nabla u|^2 + W(u) \right) dx + \frac{\sigma}{2} \int_\Omega \left| (-\Delta)^{-1/2} (u(x) - \mu) \right|^2 dx \,. \qquad (1)$$

In this formula, the real number $\mu = \int_\Omega u \, dx / |\Omega|$ denotes the average mass of u, and the nonlinear function W is a double-well potential with global minima at ± 1. For the sake of simplicity, we consider the double-well potential $W(u) = (u^2 - 1)^2/4$, although this is not essential. Note that the energy functional (1) is just the standard van der Waals free energy, but with an additional nonlocal term which involves the square root of the inverse Laplacian on a space with zero total mass. While there are different ways in which one can associate gradient-like dynamics to (1), the diblock copolymer model in its standard form considers the evolution equation

$$u_t = -\Delta \left(\epsilon^2 \Delta u - W'(u) \right) - \sigma (u - \mu) \quad \text{in } \Omega \,, \qquad (2)$$

$$\mu = \frac{1}{|\Omega|} \int_\Omega u(x) \, dx \,, \qquad \text{and} \qquad \frac{\partial u}{\partial \nu} = \frac{\partial \Delta u}{\partial \nu} = 0 \quad \text{on } \partial\Omega \,,$$

which is based on the gradient in the H^{-1}-topology and uses homogeneous Neumann boundary conditions for both u and Δu. In this formulation, the value of u describes the local material composition in the following way. Values of the phase variable $u(t,x)$ which are close to $+1$ are interpreted as only component A being present at a point $x \in \Omega$ and at time $t \geq 0$, and the value -1 indicates that only component B is present; values in between correspond to mixtures of the two components, with zero representing an equal mixture. The parameter μ denotes the average mass of the mixture, and the two remaining parameters $\epsilon > 0$ and $\sigma \geq 0$ are dimensionless interaction lengths. Informally, $\epsilon > 0$ being small corresponds to short range repulsions being strong, inducing a strong compulsion to separate, while σ being large represents strong long range chain elasticity forces, inducing a strong compulsion to hold together. Notice that for $\sigma = 0$ the diblock copolymer model (2) reduces to the celebrated Cahn-Hilliard equation [6], which serves as a basic model for the phase separation phenomena spinodal decomposition [33, 34, 49, 50, 54] and nucleation [2, 3, 14].

While deterministic phase-field models such as (2) can accurately describe a wide variety of phenomena, they do completely ignore thermal fluctuations which are present in any material. In the context of the Cahn-Hilliard equation, this shortcoming was addressed by Cook [11], who extended the deterministic model by incorporating random thermal fluctuations. A similar approach can also be used for the diblock copolymer model (2), and this leads to the modified evolution equation

$$u_t = -\Delta\left(\epsilon^2 \Delta u - W'(u)\right) - \sigma(u - \mu) + \eta \cdot \xi(t, x) \,, \tag{3}$$

where the additive noise term $\xi(t,x)$ is usually chosen as space-time white noise or colored noise, and the factor η describes the noise intensity. For more details on such processes, in particular their precise definition, construction, and properties, we refer the reader to [4].

Using the above setting, the remainder of this paper is organized as follows. In Sect. 2 we show how computational homology can be used to quantify the topological microstructure evolution during the initial phase separation from a homogeneous state in both (2) and (3). In Sect. 3 we turn to dynamical questions. For this, we restrict our attention to the case of one-dimensional domains Ω and review some recent numerical results by Johnson et al. [27], which provide insight into the typical frequency of long-term limits of the equation. We also indicate how these observations can be made rigorous using computer-assisted topological methods, and answer an open question from [27]. Finally, in Sect. 4 we outline further work in this area, including a number of open questions. We would like to emphasize that the selection of material in this article is purely based on the author's interest, and that the article is not a comprehensive account of topological methods in evolution equations. However, we hope that it can serve as a point of entry for interested researchers.

2 Homological Microstructure Analysis

This section is devoted to the study of microstructures using tools from algebraic topology. We begin by introducing homology groups and their associated Betti numbers, and then present some novel simulation results for the two-dimensional diblock copolymer model. This homological analysis will allow us to quantify the effect of noise in (3) on the microstructure evolution. Finally, we address some computational aspects of homological methods.

Homology Groups and Betti Numbers Most phase separation phenomena in materials science produce highly irregular and intriguing patterns, which often involve an element of stochasticity or randomness. For example, in Fig. 1 we show six different microstructures generated through numerical simulations of the diblock copolymer model (2). In each of these simulations, the evolution starts at a random small-amplitude perturbation of the homogeneous state $u \equiv \mu$, which models the

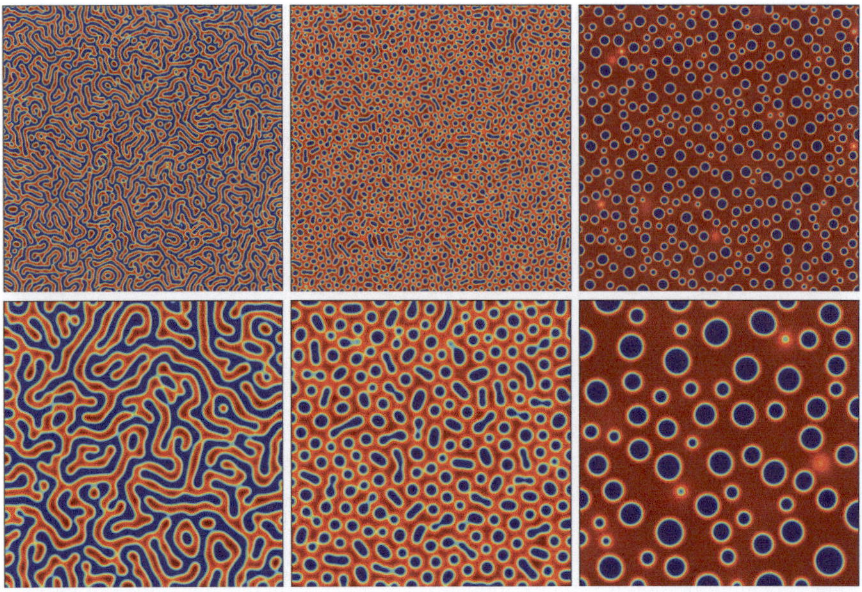

Fig. 1 Sample two-dimensional microstructures generated via numerical simulations of (2) on the domain $\Omega = (0, 1)^2$. All six cases use the parameter value $\sigma = 10$, and the solution snapshots are shown after the initial phase separation is complete and coarsening sets in. From left to right the images in each column correspond to total mass $\mu = 0$, $\mu = 0.25$, and $\mu = 0.5$, respectively. The *top row* is for $\epsilon = 0.002$, while the *bottom row* is for $\epsilon = 0.005$. The positive nodal domain $N^+(t)$ is shown in *red*, while the negative one is shown in *blue*

mixture at high temperature and in thermodynamical homogeneous equilibrium. After quenching, phase separation sets in, and the phase variable u starts to grow to values close to ± 1 throughout most of the domain. Once this initial separation of spinodal decomposition type is completed, coarsening sets in. All of the six solution snapshots in Fig. 1 are taken at that point in time. While these patterns are clearly different on a qualitative level, a quantitative metric for distinguishing them would be desirable.

Since the patterns described above do not generally exhibit any periodicity or other simplifying geometric features, we make use of algebraic topology to measure essential connectivity information in each dimension. More precisely, we use computational homology to associate a sequence of nonnegative integers called Betti numbers to each microstructure. Since $u(t,x)$ represents the relative concentration difference between the two monomer fractions at time t and location x, the simplest decomposition of the domain Ω consists of the two sets

$$N^+(t) = \{x \in \Omega \mid u(t,x) \geq m\} \quad \text{and} \quad N^-(t) = \{x \in \Omega \mid u(t,x) \leq m\} , \qquad (4)$$

where m denotes a suitable threshold value, such as the total mass $m = \mu$. In this situation, the sets $N^\pm(t)$ represent the regions in the material where one or the other monomer dominates. The sets $N^{\pm}(t)$ are called the nodal domains of u—which are often also referred to as excursion sets, especially in a stochastic context. Notice that since u is the solution of a possibly stochastic nonlinear partial differential equation, we cannot expect to have an explicit representation of the sets $N^{\pm}(t)$. However, as subsets of $\mathbb{R}^d$ the nodal domains $N^{\pm}(t)$ are topological spaces, and therefore we can consider their associated homology groups $H_p(N^{\pm}(t))$ for $p = 0, 1, 2, \ldots$. While the precise definition of homology groups can be found in [28], see also the references therein, in our context it suffices to realize that in fact

$$H_p(N^{\pm}(t)) \cong \mathbb{Z}^{\beta_p^{\pm}(t)} \quad \text{for} \quad p \in \mathbb{N}_0 ,$$

where $\mathbb{Z}$ denotes the set of integers, and $\beta_p^{\pm}(t)$ is a non-negative integer called the p-th Betti number. Thus, in our nodal domain situation the homology groups are characterized by their Betti numbers.

Each of the Betti numbers measures a different geometric property. It is shown in [28] that the 0-dimensional Betti number $\beta_0^{\pm}(t) \geq 1$ counts the number of connected components or pieces of the nodal domain $N^{\pm}(t)$. We would like to point out that the size or the shape of the components does not play a role in the value of $\beta_0^{\pm}(t)$. For example, in the right-most upper image of Fig. 1 the red set $N^+(t)$ consists of only one component, while the blue set $N^-(t)$ decomposes into $\beta_0^-(t) = 401$ pieces. Similarly, the right-most image in the second row of the figure has $\beta_0^+(t) = 1$ and $\beta_0^-(t) = 71$. In contrast, the left-most upper image gives Betti numbers $\beta_0^+(t) = 129$ and $\beta_0^-(t) = 124$, while the left-most lower image has Betti numbers $\beta_0^+(t) = 33$ and $\beta_0^-(t) = 41$. The independence of the Betti numbers from the specific shape can be either a strength or a weakness of this approach, but in cases without a priori knowledge of the specific geometry of the patterns it frequently is the former.

The Betti numbers for $p \geq 1$ measure holes in dimension p. Thus, $\beta_1^{\pm}(t)$ provides a measure of the number of tunnels in the set $N^{\pm}(t)$, though the correspondence is slightly more complicated. In two-dimensional domains, such as the ones depicted in Fig. 1, tunnels are reduced to loops which cannot be contracted to a point within the set. Thus, the lower left-most image has $\beta_1^+(t) = 11$ and $\beta_1^-(t) = 8$, while the bottom right-most image has $\beta_1^+(t) = 56$ and $\beta_1^-(t) = 0$. In the latter image, notice that even though the blue set $N^-(t)$ has many components, it does not form any holes. In contrast, the red set $N^+(t)$ has only one component, but forms 56 holes—which correspond exactly to the interior components of the blue set. We will return to this point later in this section.

While for the purposes of this paper the above-described Betti numbers suffice, their higher-dimensional analogues are important for three-dimensional microstructures. In such a structure, loops become tunnels, where again the length or width of the tunnel is irrelevant. For example, a washer has one wide but very short tunnel, while a garden hose has a narrow but long tunnel. In either case, the first Betti number is one. Moreover, for general three-dimensional domains the Betti number $\beta_2^{\pm}(t)$ equals the number of enclosed volumes or cavities within the material.

Homological Analysis of Evolving Patterns One of first applications of homology theory to the study of evolving complicated patterns was in the context of material microstructures produced by the Cahn-Hilliard models. It has been shown in recent years that the dynamics of these patterns, both in the deterministic and in the stochastic setting, is to a large extent determined by unexpectedly linear effects [5, 15, 49, 50, 54]—and on a purely qualitative level the microstructures produced by the deterministic Cahn-Hilliard model [6] and the stochastic Cahn-Hilliard-Cook model [11] appear to be similar. Nevertheless, it was demonstrated in [22] that Betti number metrics can be used to identify the suitability of these evolution equation models in the context of isothermal and mass-conserving phase separation on two-dimensional domains. In this paper, computational homology was used to quantify differences in the microstructure evolution between the deterministic and the stochastic models, which effectively ruled out the applicability of the deterministic Cahn-Hilliard model during spinodal decomposition due to the experimental results in [26]. Moreover, we could shed light on the effect of the domain boundary on the phase separation process, and isolate bulk effects from boundary effects. It is important to note that these results do require the use of the Betti numbers. As was pointed out in [22], the frequently used Euler characteristic cannot isolate the bulk information, see also our discussion below.

To illustrate a very basic homological analysis of pattern evolution, we now present some new computational results for the diblock copolymer model (3) with additive stochastic term. This equation is a stochastic evolution equation over a probability space $(F, \mathcal{F}, \mathbb{P})$, and even in the deterministic special case $\eta = 0$ it contains an element of randomness through the choice of random initial conditions over this probability space. In both situations, the nodal domains are actually random sets $N^{\pm}(t, \omega)$, where $\omega \in F$, which are given as nodal domains of a random field $u(t, x, \omega)$. Since we are interested in typical results for the underlying model, we determine the averaged Betti number evolution curves from an ensemble of 1024 runs with random small-amplitude perturbations of the homogeneous state $u \equiv \mu$. More precisely, we consider the domain $\Omega = (0, 1)^2$ and the parameter values $\epsilon = 0.005$, $\sigma = 10$, and $\mu = 0$, as well as the four noise intensities $\eta = 0$, $\eta = 0.005$, $\eta = 0.010$, and $\eta = 0.020$, respectively. The resulting averaged Betti number evolution curves for the positive nodal domains $N^+(t, \omega)$ are shown in Fig. 2. The left image depicts the average number of components, i.e., the averaged 0-dimensional Betti numbers $\beta_0^+(t)$, and the right image is for the 1-dimensional Betti numbers $\beta_1^+(t)$. We would like to point out that studying the mean of the Betti number evolution curves could certainly be supplemented by a more in-depth statistical treatment, i.e., one could also study the variance or even the distribution of the Betti number values. In the above situation, however, the averaged Betti numbers provide the central information, as the observed variances are fairly small, and the distributions are concentrated close to the mean.

The averaged Betti number evolution curves demonstrate that on the level of the produced microstructures, the behavior of the stochastic diblock copolymer model depends on the noise intensity. For the deterministic special case $\eta = 0$, one observes non-monotone pattern complexity evolution in dimensions zero and

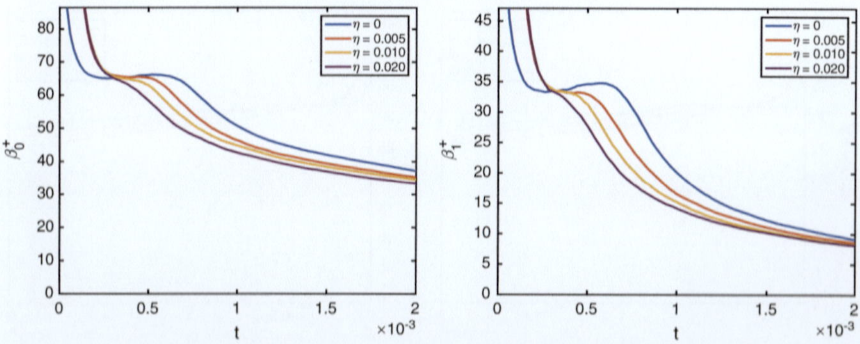

Fig. 2 Averaged Betti number evolution curves for the diblock copolymer model (3) on the square domain $\Omega = (0, 1)^2$. All simulations are for parameter values $\epsilon = 0.005$, $\sigma = 10$, and $\mu = 0$, averaged over a random ensemble consisting of 1024 independent runs. The images show the curves for the positive nodal domain, the ones for the negative nodal domain are similar and therefore omitted. The *left image* depicts the number of components, i.e., the 0-dimensional Betti number, and the *right image* is for the 1-dimensional Betti number. In each of the images, four curves are depicted which are for noise intensities $\eta = 0$, $\eta = 0.005$, $\eta = 0.010$, and $\eta = 0.020$, respectively

one, and as the noise intensity increases, this non-monotonicity starts to weaken, and eventually disappears completely. While the curves in Fig. 2 are determined from numerical simulations, similar curves can in principle also be determined from experiments, as was done in the Cahn-Hilliard setting in [25, 26, 35]. In this way, one can perform a quantitative comparison between experiment and phenomenological model, and potentially even identify model parameters.

While the use of Betti numbers in materials science is a very recent occurrence, other topological invariants have been used in the past as well. One of the most popular among these is the Euler characteristic χ, which can be easily computed from local information, since it is an additive set measure. Results from algebraic topology show that the Euler characteristic is related to the Betti numbers via the formula

$$\chi = \beta_0 - \beta_1 + \beta_2 - \beta_3 + \ldots = \sum_{p=0}^{\infty} (-1)^p \beta_p ,$$

and in certain situations it can be used to distinguish microstructures. However, in the situation of Fig. 2 its information is very limited.

To explain this, we consider again the microstructure in the right-most image of the second row in Fig. 1. It was already mentioned that the nodal domains of this solution snapshot have Betti numbers $\beta_0^+(t) = 1$, $\beta_1^+(t) = 56$, $\beta_0^-(t) = 71$, and $\beta_1^-(t) = 0$. Moreover, loops in the nodal domain $N^{\pm}(t)$ correspond to internal components in $N^{\mp}(t)$, i.e., to components which do not touch the boundary of the domain. This clearly opens up the possibility of distinguishing between interior

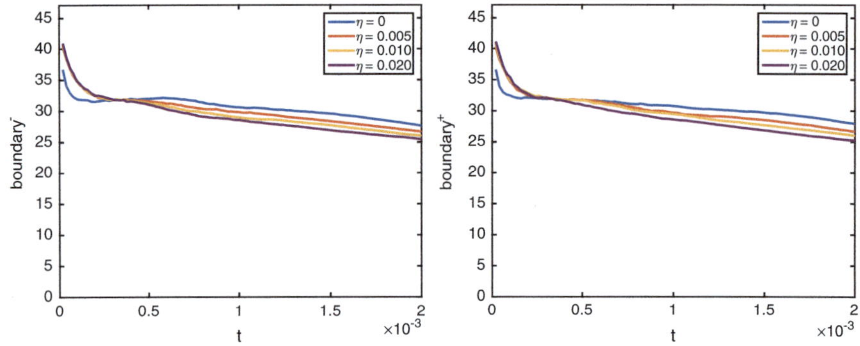

Fig. 3 Averaged evolution curves for the number of boundary components of the patterns which were used to generate Fig. 2. The *left* and *right images* are for the positive and negative nodal domains, respectively. Notice that these curves more or less agree, i.e., only the interior or bulk behavior allows one to distinguish the noise amplitude in the model (3)

and boundary components of the microstructure, and allows one to separate bulk behavior from boundary behavior. In fact, if we denote the number of internal components of $N^{\pm}(t)$ by $\beta^{\pm}_{\text{int},0}(t)$ and the number of components touching the boundary of Ω by $\beta^{\pm}_{\text{bdy},0}(t)$, then we have

$$\beta^{+}_{\text{int},0}(t) = \beta^{-}_{1}(t) \quad \text{and} \quad \beta^{+}_{\text{bdy},0}(t) = \beta^{+}_{0}(t) - \beta^{-}_{1}(t) ,$$

and similarly for their $N^{-}(t)$-counterparts. In the situation of the right-most image in the second row of Fig. 1 this implies $\beta^{+}_{\text{int},0}(t) = 0$, $\beta^{+}_{\text{bdy},0}(t) = 1$, $\beta^{-}_{\text{int},0}(t) = 56$, and $\beta^{-}_{\text{bdy},0}(t) = 15$, and this can easily be verified by direct counting.

We now finally return to the evolution curves in Fig. 2. According to the above formulas, the right image depicts the averaged evolution curve of the interior components of $N^{-}(t)$. But what about the boundary components? Their corresponding averaged evolution curves are shown in Fig. 3, and they clearly do not exhibit the characteristic change from non-monotone to monotone behavior as η increases from zero. In other words, only the Betti numbers detect this behavior. Furthermore, notice that due to $\mu = 0$ and $W(u) = (u^2 - 1)^2/4$, the ensemble averages of $\beta_1(N^{+}(t))$ and $\beta_1(N^{-}(t))$ have to be equal. Thus, the averaged number $\beta^{+}_{\text{bdy},0}(t)$ of boundary components equals in fact the averaged Euler characteristic of the set $N^{+}(t)$, i.e., the averaged Euler characteristic cannot be used to distinguish different noise intensities. An analogous observation can be made for the averaged Euler characteristic of $N^{-}(t)$.

As we mentioned earlier, the simulations shown in Figs. 2 and 3 provide a first glimpse into the microstructure complexity evolution of the diblock copolymer equation during the initial phase separation stage. How do they compare to the Cahn-Hilliard situation, i.e., the case $\sigma = 0$ of a purely local energy (1)? It turns out that on a quantitative level, the above data appears to be unchanged from the results of [22].

While this might seem anticlimactic, it does have some important consequences. On large domains Ω, the initial stage of phase separation lays down the overall structure of the pattern. Our results indicate that this structure is not significantly affected by the nonlocal term in the energy, and that all the differences that can be observed between the classical Cahn-Hilliard model and the diblock copolymer equation have to be due to global effects. Some of these will be addressed in the next section.

The discussions so far only provide a small and simple example of the potential of homological pattern analysis. These methods have also been used in quite a few studies during the last decade, but we only mention some of them. In the context of the diblock copolymer model, Teramoto and Nishiura have used computational homology to study the dynamics of morphology transition in [53]. Krishan et al. [30] have been able to establish the breakdown of the Boussinesq approximation in numerical simulations of convection. In the context of granular materials, Kramár et al. [29] have studied force networks. Finally, Fuller et al. [56] have linked internal stresses in polycrystals to the level of grain misorientation, based on suitably thresholded data that can be classified via homology. More recently, researchers have begun to use persistent homology [19] to address the problem of noisy data, a topic which lies beyond the scope of this article.

Computational Aspects and Pattern Validation We close this section with a brief discussion of computational aspects of homological analysis. One of the significant advantages of homology over other topological invariants, such as fundamental groups, has always been its computability. In principle, homology groups can be determined using the Smith normal form algorithm, which is for example explained in [40]. For large data sets, however, and "large" here would even include the images contained in Fig. 1, its computational complexity is unacceptable. Over the last few years, a number of algorithms have been developed to reduce the computational time. All of these algorithms aim at decreasing the size of the underlying complex without changing its homology, via so-called reduction steps. This can be done in a variety of ways, and is often connected to the actual discrete representation of the topological space. In the context of pattern analysis described in this section, cubical complexes as described in [28] are the primary choice, but the use of simplicial complexes as described in [19, 40] is also possible. The complex reductions can be performed using free-face collapses [28] or coreductions [16, 38]. More recently, algorithms have been developed which use discrete Morse theory for the reduction step [24], or which perform coreductions in the persistent homology context [39]. At the time of writing, several high-performance computational packages for homological analysis are available, for example CHomP [9] and RedHom [45], as well as Perseus [41] in the context of persistence.

The methods described in the previous paragraph enable us to determine the homology of a finite representation of the nodal domains $N^{\pm}(t)$. However, since this process necessarily involves a discretization step, errors can and sometimes will occur. In order to minimize and control these approximation errors, a number of approaches have been developed. One approach is based on the already mentioned fact that in many applications the nodal domains in question are random sets. In this case, one would like to be able to obtain a probabilistic statement which provides a correctness probability of the nodal domain approximation as a function of the discretization. In the context of Gaussian random fields, this has been accomplished in [12, 36, 37], and it leads to generally sharp a priori estimates for the necessary discretization size as a function of error tolerance. In fact, it even provides guidance as to were the random fields have to be sampled to minimize the approximation errors. Another, more computational, approach was developed in [10, 13]. In these papers, rigorous computations using interval arithmetic are used to create an adaptive rectangular decomposition of the domain Ω of a function u in such a way that each of the rectangles R in the final grid has the property that the function values of u on the corners of R correctly describe the location of the nodal line within R in the following sense. If the function values at the corners are all positive, then the function u is positive on all of R; if exactly one corner has a negative function value, while the remaining three have positive u-values, then the zero set of u in R consists of a simple non-interesting curve which originates on one edge of the square and ends at another edge; etc. In other words, the method provides a computer-assisted proof for the correctness of the generated adaptive discretization, and the final adaptive cubical grid can then easily be used to compute the correct homology of the nodal domains. The algorithm in [10] uses randomized subdivision techniques to avoid grid alignment issues in the study of evolution equations.

3 Long-Term Dynamics and Partial Attractor Structure

The methods of the previous section were aimed at describing and quantifying essential aspects of the microstructures which are created by phase-field models such as the diblock copolymer equation (2). With this section, we turn our attention to answering underlying dynamical questions. In particular, we demonstrate through numerical results that in many situations one can explain specific questions regarding the dynamical behavior by considering specific parts of the global attractor associated with a given model. To focus our discussion, we only consider one example, i.e., multistability in the diblock copolymer model.

Equilibrium Bifurcation in the Diblock Copolymer Model As mentioned before, the diblock copolymer model (2) is a gradient system with respect to the energy $E_{\epsilon,\sigma}$ defined in (1). This implies that if $u(t)$ is any solution of the partial differential equation, the real-valued function $t \mapsto E_{\epsilon,\sigma}[t]$ is monotonically decreasing. This fact can readily be exploited for obtaining a rough intuition about the phase separation process. Since the function $W(u) = (u^2 - 1)^2/4$ has two equal global minima at ± 1, the second part in the first integral of (1) becomes small if the function values of u separate towards ± 1, and since the first part of this integral penalizes large gradients, this separation preferably occurs on fairly large subsets of the domain Ω. In contrast, the second integral is a nonlocal term which includes long-range interactions, and therefore the regions formed during the separation process cannot be too large either. The competition of these phenomena ultimately leads to the complicated patterns which were observed in the previous section.

Beyond its heuristic implications, the energy associated with the diblock copolymer model can be used to show that the equation has a compact global attractor $\mathcal{A}$, see for example [52]. Loosely speaking, a compact attractor of an evolution equation is a bounded set which attracts all solutions $u(t)$ as $t \to \infty$, i.e., it describes the long-term dynamics of the complete model. In many cases, such as for example the diblock copolymer model, the attractor is in fact of finite dimension, and therefore the essential dynamics can be determined from a suitable finite-dimensional projection of the infinite-dimensional flow. One can show that generally, a compact global attractor consists of finitely many equilibrium solutions[1] of the underlying evolution equation, i.e., solutions which do not change in time. Between these stationary states, heteroclinic solutions describe transitional behavior. These are nonconstant solutions $u(t)$ which exist for all times t, and which converge to different equilibrium solutions as $t \to \pm\infty$. In other words, if one can describe the complete set of equilibria, as well as the set of heteroclinic connections between them, then the global long-term behavior of the equation is known.

To focus our discussion further, we now consider the diblock copolymer model (2) on the one-dimensional domain $\Omega = (0, 1)$, and begin by trying to shed some light on its equilibrium set. Since stationary solutions of this model have to satisfy the nonlinear problem

$$-\Delta\left(\epsilon^2 \Delta u - W'(u)\right) - \sigma(u - \mu) = 0 \quad \text{in } \Omega , \tag{5}$$

subject to the mass constraint $\mu = \int_\Omega u(x)\,dx/|\Omega|$ and homogeneous Neumann boundary conditions for both u and Δu, this reduces to solving a nonlinear elliptic equation. For the special case $\sigma = 0$ of the Cahn-Hilliard equation, this has been accomplished in [23]. In this paper, the complete equilibrium set is described for the domain $\Omega = (0, 1)$, and for varying mass values μ. Specifically for $\mu = 0$, their results can be described as follows. One can easily see that for all values of the parameter $\epsilon > 0$, the constant function $u(x) \equiv 0$ is a solution of (5) for $\sigma = 0$. As ϵ decreases towards zero, new nontrivial solutions start to appear close to this trivial zero solution. These new solutions bifurcate from the trivial solution at the parameter values $1/\epsilon^2 = k^2\pi^2$, where $k \in \mathbb{N}$. For every k-value, two nonconstant solutions appear, which qualitatively look like the function $\phi_k(x) = \sqrt{2}\cos k\pi x$. These solutions grow in amplitude as ϵ decreases, ultimately reaching a maximum norm close to one. We would like to point out that the functions ϕ_k for $k \in \mathbb{N}$ form a complete $L^2(0, 1)$-orthonormal set in the space of square-integrable functions with mean zero. This will be useful later on in the context of computer-assisted proofs.

---

[1] We assume for the purposes of this survey that equilibrium solutions are generally hyperbolic, i.e., the linearization of the evolution equation at the stationary solutions has no zero eigenvalue.

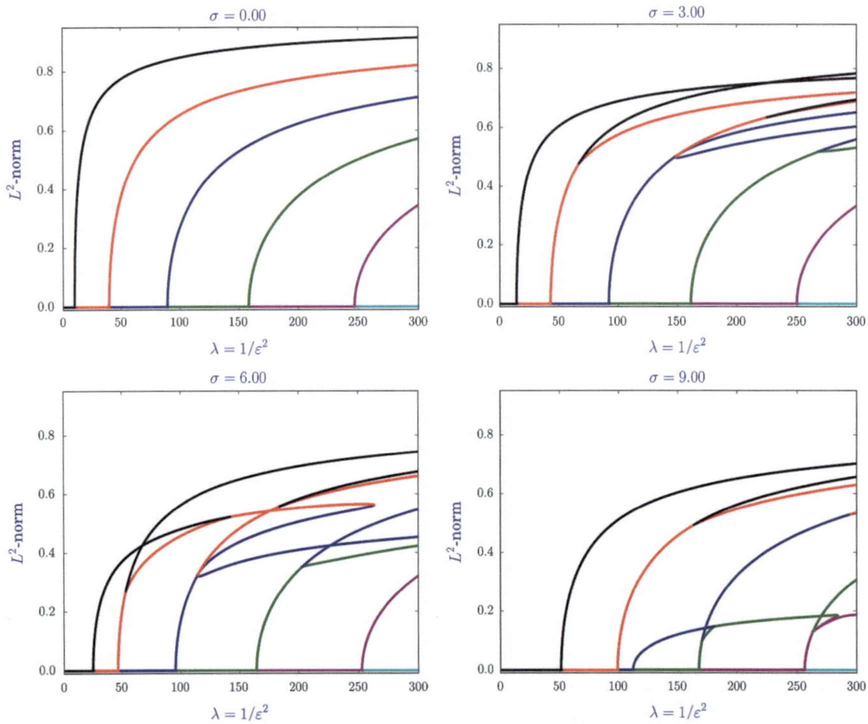

Fig. 4 Sample equilibrium bifurcation diagrams for the diblock copolymer model (2) on the one-dimensional domain $\Omega = (0, 1)$ and for total mass $\mu = 0$. From top left to bottom right the bifurcation diagrams are for $\sigma = 0, 3, 6$, and 9, respectively. In each diagram, the vertical axis measures the $L^2(0, 1)$-norm of the solutions, and the horizontal axis uses the parameter $\lambda = 1/\epsilon^2$. The solution branches are color-coded by the Morse index of the solutions, and *black*, *red*, *blue*, *green*, *magenta*, and *cyan* correspond to indices 0, 1, 2, 3, 4, and 5, respectively

This behavior is depicted in the top left image of Fig. 4. This image is a so-called bifurcation diagram, which encapsulates the most important information about the set of equilibria for the Cahn-Hilliard model. In order to make the presentation clearer, we describe the solution set using the new parameter $\lambda = 1/\epsilon^2$, i.e., studying the transition $\epsilon \to 0$ is equivalent to $\lambda \to \infty$. While the horizontal axis in the image corresponds to λ, the vertical axis measures the $L^2(0, 1)$-norm of an equilibrium solution u. Each colored point in the image corresponds to at least one solution of (5) for $\sigma = 0$ and $\mu = 0$. Points on the horizontal line $\|u\|_{L^2(0,1)} = 0$ are clearly the trivial solutions, while points on the emanating curves are nontrivial ones. Moreover, each point with $\|u\|_{L^2(0,1)} > 0$ gives rise to two solutions, since for the chosen parameter values, if u is a solution of (5), then so is $-u$. Thus, the bifurcation diagram shown in the top left of Fig. 4 indicates that for every $k^2\pi^2 < \lambda \leq (k+1)^2\pi^2$, the Cahn-Hilliard model with mass $\mu = 0$ has exactly $2k + 1$ equilibrium solutions.

How do these rigorous results change when we consider the diblock copolymer case $\sigma > 0$? At first glance, one would assume that any changes should be minor, since the diblock copolymer model arises from the Cahn-Hilliard model through a regular perturbation at $\sigma = 0$. Yet, nothing could be further from the truth. In the remaining images of Fig. 4, we present bifurcation diagrams for increasing values of σ, all of which were computed numerically using the path-following package AUTO [17], see also the description in [27]. As in the Cahn-Hilliard case, every point on a curve corresponds to at least one equilibrium solution. In fact, due to symmetries in the equation, some points correspond to even four solutions. Note, however, that the bifurcation structure of the diblock copolymer case is considerably more complicated than in the Cahn-Hilliard setting, with many secondary bifurcations.

It was shown in [27] that the basic qualitative changes in the bifurcation diagrams of (5) for $\mu = 0$ as σ increases from zero can be uncovered by a local bifurcation analysis. As σ increases, the branches emanating from the trivial solution line move to the right, but at varying speeds. In fact, the 1-branch, i.e., the branch on which solutions have the same shape as $\phi_1(x) = \sqrt{2}\cos \pi x$, moves quickly to the right, and interacts will all the subsequent k-branches for $k \geq 2$, before it moves off to infinity and ceases to exist for $\sigma \geq \pi^2$. Next, the 2-branch moves to the right and interacts for the k-branches for $k \geq 3$, before it moves off to infinity and ceases to exist for $\sigma \geq 2^2\pi^2$, and so forth. All of the local branch interactions can be analyzed rigorously using a Lyapunov-Schmidt reduction at two-dimensional kernels [27], and this shows that every local interaction involves a secondary bifurcation point on one of the participating branches. In other words, while the original Cahn-Hilliard model involves no secondary bifurcations at all, the diblock copolymer model has an abundance.

It turns out that these secondary bifurcation points are extremely important for the long-term behavior of solutions to (2), and this will be the subject of the next subsection. Note, however, that all of the bifurcation diagrams shown in Fig. 4 were computed numerically. While the first one has been established rigorously in [23], their methods do not apply to the diblock copolymer case. In fact, to this day there are no rigorous classical mathematical results which verify the remaining bifurcation diagrams. While we strongly believe that such results are basically out of reach, computer-assisted proofs can be obtained, and will in fact be presented below using topological fixed point arguments.

Multistability in the Diblock Copolymer Model The bifurcation diagrams shown in Fig. 4 contain significantly more information than discussed so far. While these diagrams indicate the occurrence of equilibrium solutions as well as their size, recall that our main interest in stationary solutions stems from dynamical questions. The equilibria are the fundamental building blocks of the global attractor, and the latter encapsulates the long-term behavior of solutions. It is clear from their definition that if we start the evolution of (2) exactly at one of the equilibria, then we will stay there for all times. If, however, one starts only close by, one generally is driven away from the equilibrium solution. In fact, after following a heteroclinic solution one converges to another stationary state. This instability can be quantified using the concept of index. Roughly speaking, an equilibrium solution u_0 has index ℓ, if at u_0 there are ℓ linearly independent directions along which solutions starting close to u_0 can be driven away.[2] Thus, if an equilibrium u_0 has index $\ell \geq 1$, generic perturbations of u_0 lead to solutions which diverge from u_0. For index zero, there are no unstable directions, and one can show that all solutions starting sufficiently close to the equilibrium are attracted to it.

In the images of Fig. 4, the stability of each equilibrium is indicated by the color of the branch, as described in the figure caption. This shows for example that as λ increases, the index of the trivial solution $u \equiv 0$ increases from zero in steps of one. Moreover, in the case of the Cahn-Hilliard model $\sigma = 0$, only the equilibrium solutions on the 1-branch are stable. In other words, one would expect that almost all solutions of the Cahn-Hilliard model will converge to these stable states. The situation is completely different in the diblock copolymer case $\sigma > 0$. As one can see from the remaining bifurcation diagrams in Fig. 4, multiple stable states exist. All of these stable states are in fact local minimizers of the energy $E_{\epsilon,\sigma}$ defined in (1), and some will be global minimizers. In view of this discussion it is natural to wonder where typical solutions of the diblock copolymer model (2) end up if they start close to the homogeneous state $u \equiv \mu$. Specifically, will generic solution converge to global or local minimizers? This question was studied numerically in [27], and we now briefly describe these results.

A considerable part of the theoretical work on stationary solutions for the diblock copolymer model has concentrated on global minimizers of $E_{\epsilon,\sigma}$. These studies have led to a large number of results establishing the existence of global minimizers on a variety of domains Ω and in various spatial dimensions. For us, a result due to Ren and Wei [46] is of particular interest. For the one-dimensional domain $\Omega = (0, 1)$, as well as for total mass $\mu = 0$ and $\sigma > 0$, they show that the global minimizer of (1) is generally uniquely determined up to multiplication by -1, and is in fact periodic with minimal period $P^{\epsilon,\sigma}$ which satisfies

$$P^{\epsilon,\sigma} = \left(96\sqrt{2} \cdot \int_{-1}^{1} \sqrt{W(s)}\, ds \right)^{1/3} \cdot \left(\frac{\epsilon}{\sigma} \right)^{1/3} + O\left(\epsilon^{2/3}\right) \quad \text{for} \quad \epsilon \to 0 \,, \tag{6}$$

where $\int_{-1}^{1} \sqrt{W(s)}\, ds = 2/3$ for our nonlinearity $W(u) = (u^2 - 1)^2/4$. This result implies that asymptotically for large integers $k \in \mathbb{N}$, the global energy minimizer has a wave number k, i.e., it is qualitatively of the same form as $\phi_k(x) = \sqrt{2}\cos k\pi x$, if the parameters ϵ and σ satisfy

$$\frac{\sigma}{\epsilon} \approx 96\sqrt{2} \cdot \frac{2}{3} \cdot \left(\frac{k}{2} \right)^3 = 8\sqrt{2} \cdot k^3 \,, \tag{7}$$

[2]We assume again that the equilibrium is hyperbolic, which is true away from the bifurcation points.

since in this case the minimal period has to be $2/k$. In other words, for small values of $\epsilon > 0$ the global minimizers of the diblock copolymer energy exhibit fine structure, i.e., periodic behavior with large wave number k. Moreover, it was shown in Nishiura, Y., Ohnishi, I. (Rugged landscape with fine structure, Unpublished, preprint) that in the limit $\epsilon \to 0$ the number of local minimizers of the energy functional $E_{\epsilon,\sigma}$ converges to infinity. We would like to point out again that these results are very different from the Cahn-Hilliard case $\sigma = 0$, in which there are only two stable equilibria with exactly one transition layer each, and which are negatives of each other.

Part of the mechanism which creates the above-mentioned stable equilibrium solutions with fine structure has already been indicated in the previous subsection. As σ increases from zero, the k-branches in the bifurcation diagram move to the right and disappear completely for $\sigma \geq k^2\pi^2$. Recall again that by a k-branch we mean the branch bifurcating from the trivial solution with equilibria whose shape is as $\phi_k(x) = \sqrt{2}\cos k\pi x$. Thus, if we fix a value of $\sigma > 0$ and consider the loss of stability of the trivial solution as $\lambda = 1/\epsilon^2$ increases from zero, chances are that the first stable equilibrium branch is a k-branch for large $k \in \mathbb{N}$, i.e., it consists of equilibrium solutions with minimal period $2/k$. In contrast, in the case of the Cahn-Hilliard model the trivial solution always loses its stability to the 1-mode solution. The following simple lemma, which is taken from [27, Lemma 2], describes in detail the stability loss of the trivial equilibrium solution of (2) as λ increases from zero.

Lemma 1 (Stability Transfer from the Trivial Solution) *Consider the nonlinear problem (5) on the one-dimensional domain $\Omega = (0, 1)$ and for $W(u) = (u^2-1)^2/4$, and assume that $\mu = 0$. Let $\kappa_k = k^2\pi^2$ denote the eigenvalues of the negative Laplacian on Ω subject to homogeneous Neumann boundary conditions. Finally, for arbitrary $\sigma \geq 0$ define the positive integer $k_{\mathrm{stab}}(\sigma)$ by*

$$k_{\mathrm{stab}}(\sigma) = m \quad \textit{for all} \quad \frac{\kappa_{m-1}\kappa_m}{\kappa_{m-1} + \kappa_m} \leq \sigma < \frac{\kappa_m\kappa_{m+1}}{\kappa_m + \kappa_{m+1}}, \quad \textit{for } m \in \mathbb{N} . \tag{8}$$

Then for fixed $\sigma \geq 0$, as λ increases from zero, the trivial solution $u = 0$ loses its stability to the nontrivial solutions on the $k_{\mathrm{stab}}(\sigma)$-branch when

$$\lambda > \lambda_{\mathrm{stab}}(\sigma) = \frac{\kappa^2_{k_{\mathrm{stab}}(\sigma)}}{\kappa_{k_{\mathrm{stab}}(\sigma)} - \sigma} . \tag{9}$$

For $\lambda \leq \lambda_{\mathrm{stab}}(\sigma)$ the trivial solution is stable. Moreover, the function $\lambda_{\mathrm{stab}}(\cdot)$ is continuous on $\mathbb{R}_0^+$, even though $k_{\mathrm{stab}}(\cdot)$ obviously is not.

The statement of the lemma is visualized in Fig. 5. In the σ-λ-parameter quadrant, the diagram depicts the function λ_{stab} in black, i.e., it is the lower envelope which clearly is continuous, but not everywhere differentiable. For parameter combinations (σ, λ) below this curve, the homogeneous state $\mu = 0$ is globally stable, at points above this curve all stable solutions are nontrivial. As can be seen

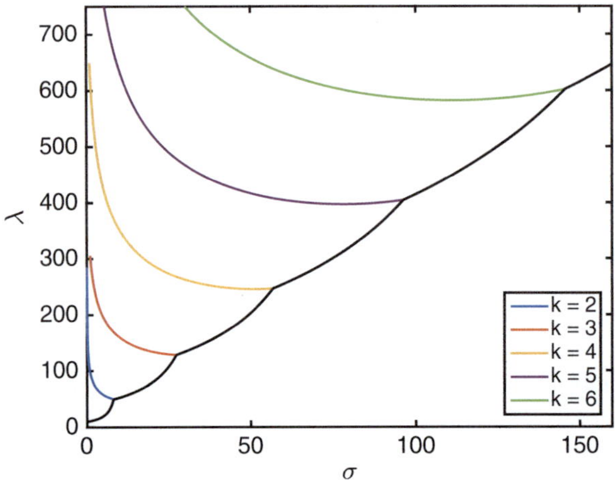

Fig. 5 Structure of typical long-term limits for solutions of (2) on $\Omega = (0, 1)$ which start close to the homogeneous state $\mu = 0$. The image partitions the σ-λ-parameter quadrant into regions with different behaviors of the long-term limit. The lower envelope is the graph of λ_{stab} in Lemma 1, i.e., for parameter values below this curve the trivial solution is the global limit. For parameter combinations in the remaining region, the periodicity of typical long-term limits is given by the wave number k, if the parameter point lies below the plotted curve for $k + 1$ and above the one for k

from the diagrams in Fig. 4, one usually has several stable equilibria with different wave numbers k at such points.

In view of the above-mentioned multistability, i.e., the occurrence of multiple local or global minimizers of the energy $E_{\epsilon,\sigma}$, which of these minimizers will be the typical long-term limit of solutions of (2) which start close to the homogeneous state $\mu = 0$? Recall that our interest in these solutions stems from the fact that they model the phase separation evolution initiated by quenching the material. In [27], extensive numerical simulations of Monte-Carlo type have shown that the first parameter quadrant in the σ-λ-plane can be partitioned into regions in such a way that for parameter combinations (σ, λ) in one of these regions most solutions of (2) originating close to the homogeneous state $\mu = 0$ lead to a long-term limit with wave number k. These regions are indicated in Fig. 5. They lie above the black lower envelope, which is the function λ_{stab}, and are delineated by the colored curves in the diagram, which are indexed by the wave numbers k. If a parameter combination lies below the plotted curve for $k + 1$ and above the one for k, then the long-term limit of most solutions has wave number k. We would like to point out that the delineating curves are very sharply defined, and the switch in wave numbers is evident as parameter combinations move across them. Note, however, that we are not saying that all initial conditions close to the homogeneous lead to the same long-term limit. There are certainly different periodicities possible, but (by a large margin) most lead to the periodicity indicated in Fig. 5. For more details we refer the reader to [27].

As it turns out, the delineating curves appear to be intimately connected to the bifurcation diagrams in Fig. 4. To describe this further, assume again that $\mu = 0$, and let $k \geq 2$ be an integer. It was shown rigorously in [27] that for σ-values right before the local interaction between the $(k-1)$-branch and the k-branch, there exists a secondary bifurcation point on the k-branch, which converges to the trivial solution as σ approaches $\kappa_{k-1}\kappa_k/(\kappa_{k-1}+\kappa_k)$. We call this point a stabilizing secondary bifurcation point on the k-mode branch, and it exists at least locally on a smooth curve for $\sigma < \kappa_{k-1}\kappa_k/(\kappa_{k-1}+\kappa_k)$. The stabilizing secondary bifurcation point for $k = 2$ can easily be seen in the upper right and lower left images of Fig. 4. It is the first secondary bifurcation point on the second bifurcation branch from the trivial solution, and the images show that its location in the σ-λ-parameter quadrant changes with σ. In fact, the points (σ, λ) at which the stabilizing secondary bifurcation point for $k = 2$ can be observed are precisely the points on the blue curve in Fig. 5, and similarly for the other values of k listed in the figure legend.

The discussion of the previous paragraph relates the behavior of the long-term limits of solutions of (2) which start close to the homogeneous state $\mu = 0$ to location curves of secondary bifurcation points in the bifurcation diagrams for (5). Since the latter bifurcation points are far away from the homogeneous state, unless we choose parameter combinations (σ, λ) close to the black lower envelope in Fig. 5, the sharp delineating effect of these curves demonstrated in the simulations in [27] is extremely surprising. While this issue could not be resolved in the cited paper, we will provide an explanation in the next subsection.

Before that, let us briefly indicate how the periodicity of the typical long-term limits relates to the periodicity of the global energy minimizers as described in (6) and (7). It was shown in [27] that if we rescale the location curves of the stabilizing secondary bifurcation points by the square of the associated wave number k, then these curves approach a universal curve as $k \to \infty$. In fact, we could numerically show that in this limit, the scaled stabilizing secondary bifurcation point on k-mode branch occurs along for

$$\frac{\lambda}{k^2} \approx 19.88 \cdot \left(\frac{\sigma}{k^2}\right)^{-0.2738} . \tag{10}$$

This is shown in Fig. 6, where the scaled location curves are plotted in their respective colors in a doubly logarithmic plot.

The formula in (10) provides an asymptotic estimate as to when the stable k-mode equilibrium can be expected to be the long-term limit of most solutions of the diblock copolymer equation originating close to the homogeneous state $\mu = 0$. As it turns out, this limiting periodicity has little to do with the periodicity of the global energy minimizer. According to (7), the stable solution on the k-mode branch becomes the global energy minimizer when λ, σ, and k satisfy

$$\frac{\lambda}{k^2} \approx 128 \cdot \left(\frac{\sigma}{k^2}\right)^{-2} , \tag{11}$$

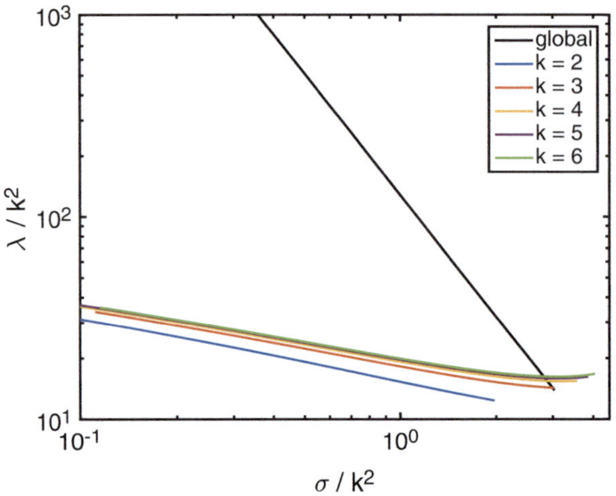

Fig. 6 Inherent scalings of the delineating curves from Fig. 5. By incorporating the wave number k into the scaling formula for these curves, one can see that they approach a single universal scaling law as k becomes large, see also (10). More precisely, if the curves from Fig. 5 are plotted in a σ/k^2-λ/k^2-coordinate system, where k is the wave number from the legend in Fig. 5, one obtains the colored curves in the above image. In addition, the black line denotes the scaled version of the parameter combinations at which the k-branch equilibrium becomes globally stable, as given by (11). Notice that for given values of σ and k, the λ-value at which the k-branch equilibrium becomes a local minimizers is significantly smaller than the λ-value at which it becomes the global minimizer

i.e., every value of k leads to the same scaled curve. This scaled curve is shown in black in Fig. 6, and it lies considerably above the curve in (10). Thus, in general the observed periodicity of the long-term limit of typical solutions of (2) which originate close to the homogeneous equilibrium is not described by the periodicity of the global energy minimizer, but rather by a local minimizer with higher periodicity. Particularly, the difference in the exponents on the right-hand sides of (10) and (11) implies that the observed periodicity will in practice be considerably larger than that of the global minimizer.

Rigorous Results via Computer-Assisted Proofs As mentioned earlier, the results described in the last subsection are all numerical in nature, and classical mathematical techniques can probably not be used to derive rigorous results. Yet, especially over the last decade, a number of approaches have been developed to obtain computer-assisted proofs using topological fixed point arguments, see for example [31, 32, 44, 51] and the references therein.

In a recent paper [55] we have adapted the approach by Plum [44] to establish equilibrium solutions for the diblock copolymer model. The method uses a topological fixed point argument to prove that in a neighborhood of a numerical approximation of the stationary state there is a unique equilibrium solution of the partial differential equation. In order to establish this, one basically needs to verify two conditions. Let u_* denote the numerical approximation, which in our case is given as a finite sum of the cosine basis functions $\phi_k(x) = \sqrt{2}\cos k\pi x$ and produced by AUTO via a spectral method. Then on the one hand we need to show that the residual

$$\rho = \left\| -\Delta \left(\Delta u_* - \lambda W'(u_*) \right) - \lambda\sigma(u_* - \mu) \right\|_Y \tag{12}$$

is sufficiently small, where Y denotes a suitable function space, such as for example $Y = H^{-3}(0,1)$. On the other hand, one needs an explicit expression for a constant $K > 0$ such that for all functions $v \in X$ the estimate

$$\|v\|_X \le K \left\| -\Delta \left(\Delta v - \lambda W''(u_*) v \right) - \lambda\sigma v \right\|_Y \tag{13}$$

holds, where X is another suitable function space, for example $X \subset H^1(0,1)$. In other words, the constant K is an upper bound on the operator norm of the inverse of the Fréchet derivative of the left-hand-side of (5) at u_*. Note that we slightly rewrote (5) by multiplying both sides with $\lambda = 1/\epsilon^2$. Both of the estimates (12) and (13) can be established rigorously using interval arithmetic, and for more details we refer the reader to [55]. For the situation in this paper, we obtain for example the following result.

Theorem 1 (Equilibrium Existence) *Consider the stationary diblock copolymer problem (5) on the domain $\Omega = (0,1)$, let $\sigma = 6$, and consider the total mass $\mu = 0$. For $\lambda = 50$ consider the two approximative solutions u_* shown in the left image of Fig. 7, and for $\lambda = 75$ consider the functions u_* shown in the right image. For each of these solutions, let the constants δ_1, δ_2, and δ_λ be as in Table 1. Then for each function u_*, there exists a uniquely determined equilibrium solution u of (5)*

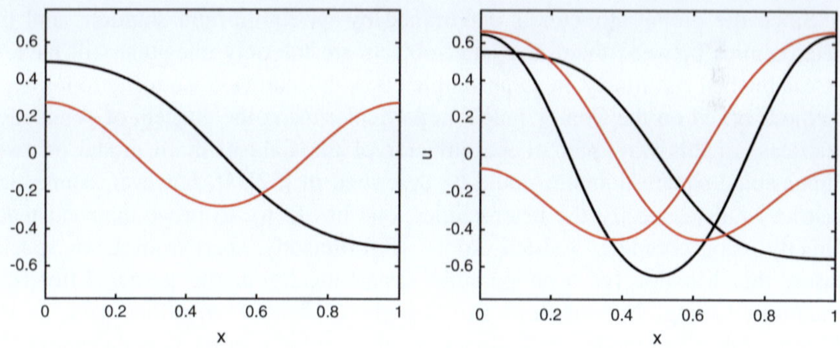

Fig. 7 Numerical approximations of equilibrium solutions for the diblock copolymer model with $\mu = 0$ and $\sigma = 6$. The *left* and *right images* correspond to λ-values 50 and 75, respectively, and the solutions are colored according to their index as in Fig. 4. If for every shown solution we also add its negative, then we obtain all nontrivial stationary solutions of the diblock copolymer equation at these λ-values

Table 1 Rigorous numerical results for the diblock copolymer model with $\mu = 0$ and $\sigma = 6$. For the λ-values 50 and 75 we consider representative solutions from each of the branches shown in the lower left diagram of Fig. 4. The equilibria are listed according to decreasing $L^2(0,1)$-norm, i.e., in the diagram they correspond to the branches from top to bottom, and the specific shape of the solutions can be found in Fig. 7

Equilibrium validation for $\mu = 0$ and $\sigma = 6$						
λ	$\|u\|_{L^2(0,1)}$	ρ	K	δ_1	δ_2	δ_λ
50	0.36999	1.330171e-14	3.9	1.030319e-13	5.699397e-04	4.425370e-04
	0.19771	2.349127e-15	10.2	4.778949e-14	2.056832e-04	2.938235e-04
75	0.46880	9.295863e-15	3.5	6.400766e-14	1.743451e-04	4.082609e-04
	0.44046	2.022536e-14	5.4	2.148780e-13	2.252343e-04	1.128654e-04
	0.38971	2.246871e-14	3.2	1.422903e-13	3.101413e-04	3.498386e-04
	0.38971	2.252944e-14	3.2	1.426749e-13	3.101413e-04	3.498386e-04

which satisfies $\|u - u_*\|_{H^1(0,1)} \le \delta_1$. *Moreover, this solution is the unique solution of (5) which has distance at most* δ_2 *from* u_*, *and it can be uniquely continued on the interval* $[\lambda - \delta_\lambda, \lambda - \delta_\lambda]$.

The computations leading to the above theorem have been implemented in Matlab using the interval arithmetic package Intlab [48]. Due to the spectral representation of our approximations u_*, extremely small residuals ρ can be achieved. These are listed in Table 1, as well as the corresponding values of K in (13). We would like to point out that these computations are basically instantaneous, and can be used to rigorously establish any equilibrium solution which is not too close to a bifurcation point—since at the bifurcation point the verification clearly has to fail due to the non-invertibility of the Fréchet derivative. In fact, our numerical experiments have shown that the regions around the bifurcation points where the method fails due to exceedingly large values of K are reasonable small.

Since the global attractor is determined by its equilibrium solutions and the heteroclinics between them, the latter objects are the only ones that still have to be established rigorously. In some simple cases this can be done using topological methods based on the Conley index, in particular using the concept of connection matrices. In this way, part of the attractor of the Cahn-Hilliard model on two-dimensional square domains could be described in [32]. If, however, connection matrices fail to specify the heteroclinics, one has to try to prove their existence directly using computer-assisted proofs. Unfortunately, apart from a few special cases, this has not yet been possible systematically in the partial differential equations setting. Nevertheless, for strongly dissipative equations such as the diblock copolymer model, finite-dimensional projections of the flow do capture the attractor dynamics, and therefore it is possible to use methods for heteroclinics in the context of ordinary differential equations. One such method has been developed in [51], and it combines topological Conley index methods with the rigorous pattern validation approach from [10] which was described at the end of the last section. For more details, we refer the reader to [51] and the references therein.

To close this section, we return to our discussion of the long-term behavior of solutions of the diblock copolymer equation (2) which originate close to the homogeneous state. While we have seen that the periodicity of the typical long-term limit changes sharply at the location curves of the stabilizing secondary bifurcation points, we have not yet been able to explain why these bifurcations, which happen far from the homogeneous state, lead to such a clearly delineated change. We now provide a heuristic explanation for this, see the schematic diagrams in Fig. 8. All of these sketches illustrate the basic dynamics of (2) for $\sigma = 6$, as the parameter λ increases from the second bifurcation point on the trivial solution line to just past the stabilizing secondary bifurcation on the 2-branch, see also the lower left image in Fig. 4. Initially, for λ-values slightly larger than the second bifurcation point from the trivial solution which occurs at $\lambda = 16\pi^4/(4\pi^2 - 6) \approx 46.6$, the situation is as shown in the upper left image of Fig. 8. The strongest unstable direction at the blue homogeneous state is leading most solutions to the globally stable black states. The precise meaning of "most" can be quantified as in [33], and is a result of local flow tangencies at the homogeneous state. As λ increases past the value $\lambda = 5\pi^2 \approx 49.3$, these local tangencies change, and this leads to the situation shown in the upper right

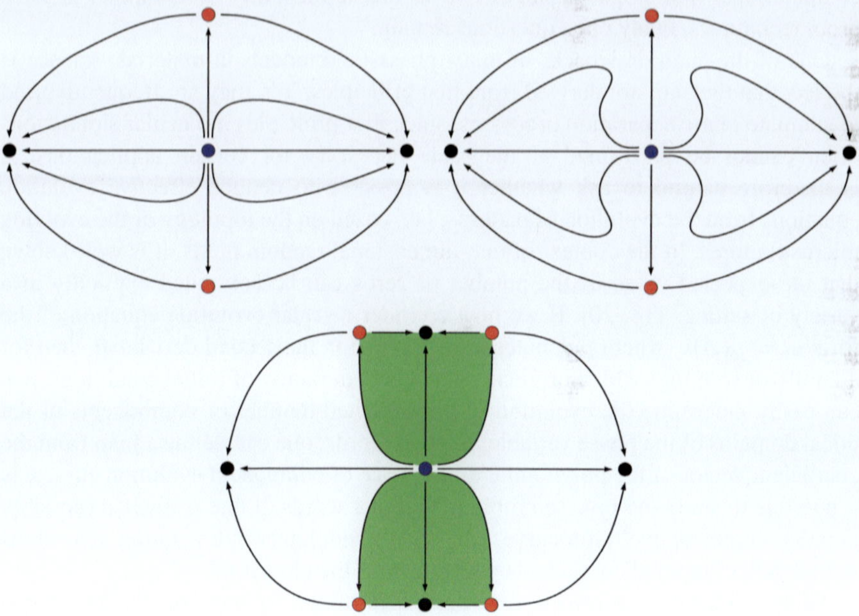

Fig. 8 Schematic explanation for the periodicity change in the long-term limits of typical diblock copolymer solutions originating near the homogeneous state, which corresponds to the *blue dot*. From top left to lower middle, the images are for increasing values of λ. While initially most solutions starting near the center are directed towards the black stable equilibria (*top left image*), the local instabilities change in order to direct most solutions to the unstable red ones (*top right image*). Once the latter equilibria undergo a secondary bifurcation, most solutions originating near the center are trapped by the newly formed stable states (*lower image*)

image of Fig. 8. Now most solutions are driven towards the red index one solutions, and from there they continue towards the same limits as before. This is the case for example at $\lambda = 50$, and the corresponding equilibria are shown in Fig. 7. Once λ passes the stabilizing secondary bifurcation, which happens at about $\lambda \approx 53.5$, the situation is as shown in the lower image of Fig. 8. The local dynamical behavior near the homogeneous state still drives most solutions towards the red index one solutions, but there they are immediately trapped by the newly created local energy minimizers. In such a way, a combination of local tangencies near the homogeneous state and global stabilizing bifurcations far away lead to the sharp delineations in Fig. 5.

4 Future Work and Open Questions

With this paper we have tried to provide a basic introduction to the use of topological methods for phase-field models in materials sciences. Despite the fact that the last decade has seen tremendous advances both in the availability and speed of computational homology packages and in the applicability of computer-assisted proof techniques, many open questions remain.

One of the main drawbacks of many phase-field models in materials science is the fact that they are not derived from first principles. Yet, they are frequently used to simulate phase separation processes, since first principles molecular simulations often cannot be performed at the scale necessary for certain applications. It is therefore natural to ask whether it is possible to deduce effective evolution equations from the evolution of patterns, i.e., based on the topology of the evolving microstructures. In the context of one-dimensional random fields, it is well-known that the expected value of the number of zeros can be computed explicitly in a variety of settings [18, 20]. If we now consider a scalar evolution equation of the form $u_t = p(\Delta)u$, where p denotes a polynomial in the second derivative, then for a family of solutions which originate at a Gaussian family of initial conditions, one can easily determine the evolution of the expected number of components of the nodal domains of the phase variable. In other words, one can define a map from the coefficient vector of the polynomial to the space of component evolution curves. Is it possible to study the inverse problem? In other words, if one is given a (possibly noisy) component evolution curve, can one use nonlinear optimization algorithms to find the polynomial coefficients which furnish the closest fit?

We have recently performed some numerical studies for numerically determined component evolution curves, which were then perturbed by additive noise. These tests have shown that the qualitative form of the dispersion polynomial can be found surprisingly reliably. Ultimately, we would like to determine whether a similar strategy could work for two-dimensional domains, and the results from our previous work on the Cahn-Hilliard equation [22] could serve as a first test case for the inverse problem. In the longer term it would be interesting to use results from molecular dynamics simulations as in [25, 26, 35] to identify the dispersion relation. Such results could be taken as a strong indicator for the correctness of a phenomenological phase-field model.

In the context of computer-assisted proofs, the main open questions are the consistent detection and verification of heteroclinic solutions and the rigorous treatment of bifurcation points. While the latter subject seems to be within reach in the near future, the situation is somewhat more complicated in the heteroclinic connection setting. The most reliable current methods use domain truncation techniques to arrive at a finite domain boundary value problem. This in turn relies on precise descriptions of local invariant manifolds, and its feasibility depends very much on the size of the index of the involved equilibria. Another strategy would be to try to establish the heteroclinics by adapting the method of [44, 55] to the unbounded domain of the heteroclinic. Regardless of these questions, it would be extremely useful to develop a general computational framework to establish the existence of equilibrium solutions, for a large class of phase-field models and nonlinearities. We hope that [55] can serve as a small first step towards this goal.

Acknowledgements The author would like to acknowledge the hospitality of the Research Institute for Mathematical Sciences (RIMS) at Kyoto University, where part of this work was done within the international research project *Toward a New Fusion Research of Mathematics and Materials Science*. The author was partially supported by National Science Foundation grants DMS-0907818, DMS-1114923, and DMS-1407087. The numerical simulations for this work were run on ARGO, a research computing cluster provided by the Office of Research Computing at George Mason University, VA (URL: http://orc.gmu.edu).

References

1. Bahiana, M., Oono, Y.: Cell dynamical system approach to block copolymers. Phys. Rev. A **41**, 6763–6771 (1990)
2. Bates, P.W., Fife, P.C.: The dynamics of nucleation for the Cahn-Hilliard equation. SIAM J. Appl. Math. **53**(4), 990–1008 (1993)
3. Blömker, D., Gawron, B., Wanner, T.: Nucleation in the one-dimensional stochastic Cahn-Hilliard model. Discret. Contin. Dyn. Syst. Ser. A **27**(1), 25–52 (2010)
4. Blömker, D., Maier-Paape, S., Wanner, T.: Phase separation in stochastic Cahn-Hilliard models. In: Miranville, A. (ed.) Mathematical Methods and Models in Phase Transitions, pp. 1–41. Nova Science Publishers, New York (2005)
5. Blömker, D., Maier-Paape, S., Wanner, T.: Second phase spinodal decomposition for the Cahn-Hilliard-Cook equation. Trans. Am. Math. Soc. **360**(1), 449–489 (2008)
6. Cahn, J.W., Hilliard, J.E.: Free energy of a nonuniform system I. Interfacial free energy. J. Chem. Phys. **28**, 258–267 (1958)
7. Choksi, R., Peletier, M.A., Williams, J.F.: On the phase diagram for microphase separation of diblock copolymers: an approach via a nonlocal Cahn-Hilliard functional. SIAM J. Appl. Math. **69**(6), 1712–1738 (2009)
8. Choksi, R., Ren, X.: On the derivation of a density functional theory for microphase separation of diblock copolymers. J. Stat. Phys. **113**, 151–76 (2003)
9. CHomP: Computational Homology Project. http://chomp.rutgers.edu (2015)
10. Cochran, G.S., Wanner, T., Dłotko, P.: A randomized subdivision algorithm for determining the topology of nodal sets. SIAM J. Sci. Comput. **35**(5), B1034–B1054 (2013)

11. Cook, H.: Brownian motion in spinodal decomposition. Acta Metall. **18**, 297–306 (1970)
12. Day, S., Kalies, W.D., Mischaikow, K., Wanner, T.: Probabilistic and numerical validation of homology computations for nodal domains. Electron. Res. Announcements Am. Math. Soc. **13**, 60–73 (2007)
13. Day, S., Kalies, W.D., Wanner, T.: Verified homology computations for nodal domains. SIAM J. Multiscale Model. Simul. **7**(4), 1695–1726 (2009)
14. Desi, J.P., Edrees, H., Price, J., Sander, E., Wanner, T.: The dynamics of nucleation in stochastic Cahn-Morral systems. SIAM J. Appl. Dyn. Syst. **10**(2), 707–743 (2011)
15. Desi, J.P., Sander, E., Wanner, T.: Complex transient patterns on the disk. Discret. Contin. Dyn. Syst. Ser. A **15**(4), 1049–1078 (2006)
16. Dłotko, P., Kaczynski, T., Mrozek, M., Wanner, T.: Coreduction homology algorithm for regular CW-complexes. Discret. Comput. Geom. **46**(2), 361–388 (2011)
17. Doedel, E.: AUTO: a program for the automatic bifurcation analysis of autonomous systems. In: Proceedings of the Tenth Manitoba Conference on Numerical Mathematics and Computing, Vol. I (Winnipeg, Man., 1980), vol. 30, pp. 265–284 (1981)
18. Edelman, A., Kostlan, E.: How many zeros of a random polynomial are real? Bull. Am. Math. Soc. **32**(1), 1–37 (1995)
19. Edelsbrunner, H., Harer, J.L.: Computational Topology. American Mathematical Society, Providence (2010)
20. Farahmand, K.: Topics in Random Polynomials. Pitman Research Notes in Mathematics, vol. 393. Longman, Harlow (1998)
21. Gameiro, M., Mischaikow, K., Kalies, W.: Topological characterization of spatial-temporal chaos. Phys. Rev. E **70**(3), 035,203, 4 (2004)
22. Gameiro, M., Mischaikow, K., Wanner, T.: Evolution of pattern complexity in the Cahn-Hilliard theory of phase separation. Acta Materialia **53**(3), 693–704 (2005)
23. Grinfeld, M., Novick-Cohen, A.: Counting stationary solutions of the Cahn-Hilliard equation by transversality arguments. Proc. R. Soc. Edinb. **125A**, 351–370 (1995)
24. Harker, S., Mischaikow, K., Mrozek, M., Nanda, V.: Discrete Morse theoretic algorithms for computing homology of complexes and maps. Found. Comput. Math. **14**(1), 151–184 (2014)
25. Hyde, J.M., Miller, M.K., Hetherington, M.G., Cerezo, A., Smith, G.D.W., Elliott, C.M.: Spinodal decomposition in Fe-Cr alloys: experimental study at the atomic level and comparison with computer models — II. Development of domain size and composition amplitude. Acta Metallurgica et Materialia **43**, 3403–3413 (1995)
26. Hyde, J.M., Miller, M.K., Hetherington, M.G., Cerezo, A., Smith, G.D.W., Elliott, C.M.: Spinodal decomposition in Fe-Cr alloys: experimental study at the atomic level and comparison with computer models — III. Development of morphology. Acta Metallurgica et Materialia **43**, 3415–3426 (1995)
27. Johnson, I., Sander, E., Wanner, T.: Branch interactions and long-term dynamics for the diblock copolymer model in one dimension. Discret. Contin. Dyn. Syst. Ser. A **33**(8), 3671–3705 (2013)
28. Kaczynski, T., Mischaikow, K., Mrozek, M.: Computational Homology. Applied Mathematical Sciences, vol. 157. Springer-Verlag, New York (2004)
29. Kramár, M., Goullet, A., Kondic, L., Mischaikow, K.: Quantifying force networks in particulate systems. Physica D **283**, 37–55 (2014)
30. Krishan, K., Gameiro, M., Mischaikow, K., Schatz, M., Kurtuldu, H., Madruga, S.: Homology and symmetry breaking in Rayleigh-Benard convection: experiments and simulations. Phys. Fluids **19**, 117,105 (2007)
31. Maier-Paape, S., Miller, U., Mischaikow, K., Wanner, T.: Rigorous numerics for the Cahn-Hilliard equation on the unit square. Revista Matematica Complutense **21**(2), 351–426 (2008)
32. Maier-Paape, S., Mischaikow, K., Wanner, T.: Structure of the attractor of the Cahn-Hilliard equation on a square. Int. J. Bifurc. Chaos **17**(4), 1221–1263 (2007)
33. Maier-Paape, S., Wanner, T.: Spinodal decomposition for the Cahn-Hilliard equation in higher dimensions. Part I: probability and wavelength estimate. Commun. Math. Phys. **195**(2), 435–464 (1998)

34. Maier-Paape, S., Wanner, T.: Spinodal decomposition for the Cahn-Hilliard equation in higher dimensions: nonlinear dynamics. Arch. Ration. Mech. Anal. **151**(3), 187–219 (2000)
35. Miller, M.K., Hyde, J.M., Hetherington, M.G., Cerezo, A., Smith, G.D.W., Elliott, C.M.: Spinodal decomposition in Fe-Cr alloys: experimental study at the atomic level and comparison with computer models — I. Introduction and methodology. Acta Metallurgica et Materialia **43**, 3385–3401 (1995)
36. Mischaikow, K., Wanner, T.: Probabilistic validation of homology computations for nodal domains. Ann. Appl. Probab. **17**(3), 980–1018 (2007)
37. Mischaikow, K., Wanner, T.: Topology-guided sampling of nonhomogeneous random processes. Ann. Appl. Probab. **20**(3), 1068–1097 (2010)
38. Mrozek, M., Batko, B.: Coreduction homology algorithm. Discret. Comput. Geom. **41**(1), 96–118 (2009)
39. Mrozek, M., Wanner, T.: Coreduction homology algorithm for inclusions and persistent homology. Comput. Math. Appl. **60**(10), 2812–2833 (2010)
40. Munkres, J.R.: Elements of Algebraic Topology. Addison-Wesley, Menlo Park (1984)
41. Nanda, V.: PERSEUS. http://www.sas.upenn.edu/~vnanda/perseus (2015)
42. Nishiura, Y., Ohnishi, I.: Some mathematical aspects of the micro-phase separation in diblock copolymers. Physica D **84**(1–2), 31–39 (1995)
43. Ohta, T., Kawasaki, K.: Equilibrium morphology of block copolymer melts. Macromolecules **19**, 2621–2632 (1986)
44. Plum, M.: Computer-assisted proofs for semilinear elliptic boundary value problems. Jpn. J. Ind. Appl. Math. **26**(2–3), 419–442 (2009)
45. RedHom: Simplicial and cubical homology. http://capd.sourceforge.net/capdRedHom (2015)
46. Ren, X., Wei, J.: On energy minimizers of the diblock copolymer problem. Interfaces Free Bound. **5**(2), 193–238 (2003)
47. Rohrer, G.S., Miller, H.M.: Topological characteristics of plane sections of polycrystals. Acta Materialia **58**(10), 3805–3814 (2010)
48. Rump, S.M.: INTLAB – INTerval LABoratory. In: Csendes, T. (ed.) Developments in Reliable Computing, pp. 77–104. Kluwer Academic Publishers, Dordrecht (1999). Http://www.ti3.tuhh.de/rump/
49. Sander, E., Wanner, T.: Monte Carlo simulations for spinodal decomposition. J. Stat. Phys. **95**(5–6), 925–948 (1999)
50. Sander, E., Wanner, T.: Unexpectedly linear behavior for the Cahn-Hilliard equation. SIAM J. Appl. Math. **60**(6), 2182–2202 (2000)
51. Stephens, T., Wanner, T.: Rigorous validation of isolating blocks for flows and their Conley indices. SIAM J. Appl. Dyn. Syst. **13**(4), 1847–1878 (2014)
52. Temam, R.: Infinite-Dimensional Dynamical Systems in Mechanics and Physics. Springer-Verlag, New York/Berlin/Heidelberg (1988)
53. Teramoto, T., Nishiura, Y.: Morphological characterization of the diblock copolymer problem with topological computation. Jpn. J. Ind. Appl. Math. **27**(2), 175–190 (2010)
54. Wanner, T.: Maximum norms of random sums and transient pattern formation. Trans. Am. Math. Soc. **356**(6), 2251–2279 (2004)
55. Wanner, T.: Computer-assisted equilibrium validation for the diblock copolymer model. Discret. Contin. Dyn. Syst. Ser. A (2016, to appear)
56. Wanner, T., Fuller Jr., E.R., Saylor, D.M.: Homological characterization of microstructure response fields in polycrystals. Acta Materialia **58**(1), 102–110 (2010)

An Energy-Consistent Model of Dislocation Dynamics in an Elastic Body

Vladimír Chalupecký and Masato Kimura

Abstract We propose an energy-consistent mathematical model for motion of dislocation curves in elastic materials using the idea of phase field model. This reveals a hidden gradient flow structure in the dislocation dynamics. The model is derived as a gradient flow for the sum of a regularized Allen-Cahn type energy in the slip plane and an elastic energy in the elastic body. The obtained model becomes a 3D-2D bulk-surface system and naturally includes the Peach-Koehler force term and the notion of dislocation core. We also derive a 2D-1D bulk-surface system for a straight screw dislocation and give some numerical examples for it.

Keywords Dislocation • Phase field model • Peach-Koehler force

1 Introduction

A dislocation or a dislocation curve is a crystallographic line defect within a crystal structure which was first studied by E. Orowan, M. Polanyi and G.I. Taylor independently in 1934. It is considered as a main mechanism of plastic deformation or yielding of the material and various material properties are studied in relation to the dislocation dynamics nowadays. See [2, 7, 9] and references therein for more details.

Some properties of dislocations are as follows. In many cases, a dislocation is a plane curve in a fixed slip plane, and is the boundary of a region shifted by the Burgers vector $b \in \mathbb{R}^3 \setminus \{0\}$. The Burgers vector is tangential to the slip plane and coincides with a translation vector of the crystal structure. Although end points often appear on the material surface or at other defects inside in a real crystal material, it is known from a topological argument that a dislocation curve cannot have an

V. Chalupecký (✉)
Fujitsu Limited, 1-17-25 Shinkamata, Ota-ku, Tokyo 144-8588, Japan
e-mail: chalupecky@jp.fujitsu.com; vladimir.chalupecky@gmail.com

M. Kimura
Faculty of Mathematics and Physics, Kanazawa University, Kakuma, Kanazawa 920-1192, Japan
e-mail: mkimura@se.kanazawa-u.ac.jp

Y. Nishiura, M. Kotani (eds.), *Mathematical Challenges in a New Phase of Materials Science*, Springer Proceedings in Mathematics & Statistics 166,
DOI 10.1007/978-4-431-56104-0_3

end point inside of the crystal lattice. Its typical length is around 10^{-6} m, typical thickness is around 10^{-9} m, and typical velocity is 10^{-6}–10^{2} m/s.

A dislocation generates a stress field around it by deforming the crystal structure, and interacts with other dislocations, defects, and far field conditions through the stress field. The virtual force from the stress field acting on the dislocation is called the Peach-Koehler force [2].

There are some mathematical models of dislocations (see [1, 2] and references therein), however, the following mathematical difficulties exist. The displacement u has a jump along the dislocation curve and it has infinite elastic energy ($u \notin H^1_{\mathrm{loc}}$). Therefore the elasticity equation is valid only outside of the dislocation core of radius about $5|b|$, where $|b|$ corresponds to the interatomic spacing.

In this paper, we construct an energy-consistent model in a mathematically clear way by means of the idea of the phase field [3]. We suppose a quasi-stationary condition, which means that the small deformation and the stress field of the material are described by the static linear elasticity equations. In other words, they are given by a minimum energy state of a suitable elastic energy. Our phase field model for the dynamics of a dislocation will be derived as a gradient flow of a total energy including the elastic one, as an analogy to the Allen-Cahn equation [4, 6].

We denote a dislocation curve by $\Gamma(t)$. In this paper, we assume the dislocation curve $\Gamma(t)$ is a smooth Jordan curve in a plane. The plane which includes the dislocation curve is called the slip plane. The crystal lattice has a defect along the dislocation curve and the defect is represented by the Burgers vector $b \in \mathbb{R}^3 \setminus \{0\}$. We suppose that b is tangential to the slip plane. Let $\tau \in \mathbb{R}^3$ be a counterclockwise tangential unit vector of the dislocation curve $\Gamma(t)$ and let $n \in \mathbb{R}^3$ be one of the two possible directions of the unit normal vector of $\Gamma(t)$ tangential to the slip plane. We call the directions n and $-n$ "outward" and "inward", respectively, in this paper. We choose a unit normal vector $\nu \in \mathbb{R}^3$ of the slip plane with which (n, τ, ν) is a right-handed orthonormal coordinate system. See Figs. 1 and 2 for a typical configuration.

One of the simplest mathematical models for the motion of a dislocation curve is [8]

$$V = -c\kappa + f, \qquad \text{on} \quad \Gamma(t), \tag{1}$$

where V is the outward normal velocity of $\Gamma(t)$ in the slip plane and κ is the inward signed curvature of the dislocation when considered as a plane curve. Equation (1) is called the mean curvature flow and is a typical mathematical model of phase transition. It is also known that it appears as a singular limit problem of the Allen-Cahn equation [4, 6].

The term f in (1) represents a far-field interaction through an elastic field from other dislocation curves including $\Gamma(t)$ itself, other defects, and boundary conditions. It is known that there is a virtual force $F \in \mathbb{R}^3$ acting on the dislocation curve which is given by the Peach-Koehler formula [2]

$$F = \tau \times (\sigma b), \tag{2}$$

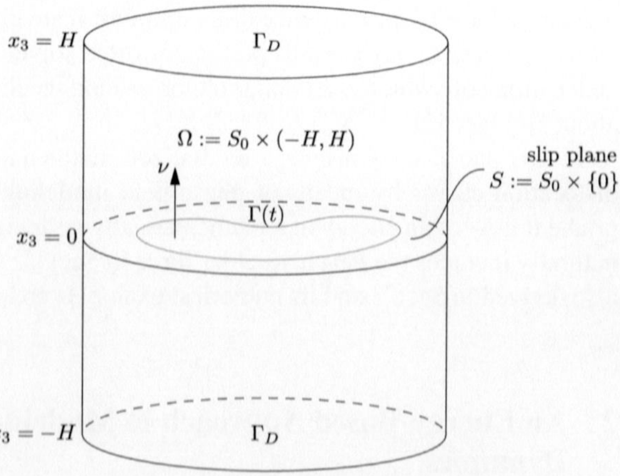

Fig. 1 A typical configuration of an elastic body with a closed dislocation curve in a slip plane

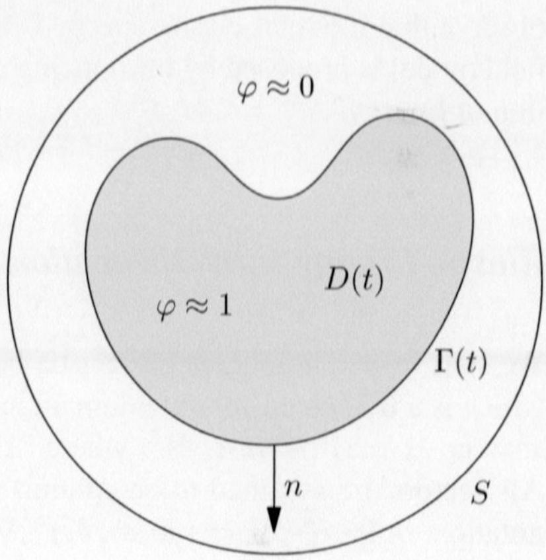

Fig. 2 A slip plane with a dislocation curve $\Gamma(t)$. The phase field φ approximately satisfies $\varphi \approx 1$ in $D(t)$ and $\varphi \approx 0$ in $S \setminus D(t)$

where $\sigma \in \mathbb{R}^{3\times 3}_{\mathrm{sym}}$ is the stress tensor field, $b \in \mathbb{R}^3$ is the Burgers vector, and $\tau \in \mathbb{R}^3$ is the unit tangential vector of $\Gamma(t)$. Then the term f in (1) is formally given by the n-direction component of the Peach-Koehler force F. Hence, we obtain

$$f = F \cdot n = (\tau \times (\sigma b)) \cdot n = (n \times \tau) \cdot (\sigma b) = \nu \cdot (\sigma b) = (\sigma \nu) \cdot b. \tag{3}$$

The Peach-Koehler formula, however, has the following mathematical problem. If we treat the dislocation as a one-dimensional curve $\Gamma(t)$, then the stress field σ mathematically has a singularity along the dislocation curve as we see in section "Elastic Energy with Dislocation" and there is no pointwise value of σ on $\Gamma(t)$. Furthermore, the stress field must have infinite energy under a most naive setting of the problem if the dislocation curve is a mathematically sharp one-dimensional object. These mathematical singularities seem to be physically regularized due to the existence of a minimum scale given by the size of atoms and lattice spacing. In physics, this problem is often solved by introducing the notion of dislocation core which is a tubular region around the dislocation line of thickness of about $5|b|$ (see [2]).

In this paper, we consider a regularized mathematical model of the motion of dislocation curves by means of phase field modeling. The model is derived as a gradient flow of an energy in a mathematically systematic way, and we show that it naturally includes the Peach-Koehler force in Sect. 2. A simplified 2D-1D model is also derived in Sect. 3 and its numerical examples are presented in Sect. 4.

2 An Energy-Based Approach to Modeling of Dislocation Dynamics

In this section, we study displacement field in an elastic body with a dislocation, and observe that the total elastic energy is infinite without any regularization. A phase field model is proposed by introducing a regularized energy in section "Regularization of Energy".

Elastic Energy with Dislocation

We start from a homogeneous elastic body without dislocation. We denote it by Ω which is a bounded Lipschitz domain in $\mathbb{R}^3$. The position vector in $\bar{\Omega}$ is denoted by $x = (x_1, x_2, x_3)^{\mathrm{T}} \in \bar{\Omega} \subset \mathbb{R}^3$, where ^{T} denotes the transpose of a vector or matrix. All vectors are assumed to be column vectors in this paper. We use the following notation: $\partial_j := \frac{\partial}{\partial x_j}$, $\nabla = (\partial_1, \partial_2, \partial_3)^{\mathrm{T}}$, $\nabla^{\mathrm{T}} u = (\partial_j u_i) \in \mathbb{R}^{3\times 3}$ and $\nabla u^{\mathrm{T}} = (\nabla^{\mathrm{T}} u)^{\mathrm{T}}$ for $u(x) \in \mathbb{R}^3$. We often use Einstein's summation convention for the space variables. For matrices $A = (a_{ij}), B = (b_{ij}) \in \mathbb{R}^{3\times 3}$, their inner product is denoted by $A : B := a_{ij} b_{ij}$.

Small deformation of the elastic body is described by a displacement field $u(x) = \big(u_1(x), u_2(x), u_3(x)\big)^{\mathrm{T}} \in \mathbb{R}^3$, the symmetric strain tensor $e[u](x) = (e_{ij}[u](x))$:

$$e[u] := \frac{1}{2}\big(\nabla^{\mathrm{T}} u + \nabla u^{\mathrm{T}}\big) \in \mathbb{R}^{3\times 3}_{\mathrm{sym}}, \quad \text{i.e.,} \quad e_{ij}[u] = \frac{1}{2}\big(\partial_j u_i + \partial_i u_j\big), \tag{4}$$

and the stress tensor $\sigma[u](x) = \big(\sigma_{ij}(x)\big) \in \mathbb{R}^{3\times 3}_{\mathrm{sym}}$:

$$\sigma[u](x) = C(x) e[u](x), \quad \text{i.e.,} \quad \sigma_{ij}(x) = c_{ijkl}(x) e_{kl}(x),$$

where $C(x) = \left(c_{ijkl}(x)\right) \in \mathbb{R}^{3\times3\times3\times3}$ is the (anisotropic) elasticity tensor with the symmetries $c_{ijkl} = c_{klij} = c_{jikl}$, $i, j, k, l \in \{1, 2, 3\}$. It should satisfy the positivity condition:

$$^{\exists}c_* > 0 \quad \text{s.t.} \quad c_{ijkl}(x)\xi_{ij}\xi_{kl} \geq c_*|\xi|^2, \qquad ^{\forall}x \in \Omega_0, \ ^{\forall}\xi \in \mathbb{R}^{3\times3}_{\mathrm{sym}}, \tag{5}$$

where $|\xi| := \sqrt{\xi_{ij}\,\xi_{ij}}$. It depends on the elastic property of the material Ω and is supposed to be given. If the material is homogeneous, the elasticity tensor should be constant $C(x) \equiv C$. From the strain-displacement relation (4), we write $\sigma[u] := C\,e[u]$.

The displacement field $u(x)$ is obtained by the following linear second order elliptic boundary value problem:

$$\begin{cases} -\partial_j\sigma_{ij}[u] = f_i(x), & i = 1, 2, 3,\ x \in \Omega, \\ u = g(x), & x \in \Gamma_{\mathrm{D}}, \\ \sigma[u]\nu = f^{\mathrm{N}}(x), & x \in \Gamma_{\mathrm{N}}. \end{cases} \tag{6}$$

where $f(x) = \left(f_1(x), f_2(x), f_3(x)\right)^{\mathrm{T}} \in \mathbb{R}^3$ is a given body force, and the first equation represents the equilibrium equations of force. The boundary $\partial\Omega$ is divided into two parts as

$$\partial\Omega = \Gamma_{\mathrm{D}} \cup \Gamma_{\mathrm{N}}, \qquad \Gamma_{\mathrm{D}} \cap \Gamma_{\mathrm{N}} = \emptyset, \qquad |\Gamma_{\mathrm{D}}| > 0,$$

where Γ_{D} is an open non-empty portion of Γ. The two dimensional area of Γ_D is denoted by $|\Gamma_{\mathrm{D}}|$. On the other hand, Γ_{N} can be empty. The outward unit normal vector to $\partial\Omega$ at $x \in \partial\Omega$ is denoted by $\nu(x) \in \mathbb{R}^3$. The displacement on Γ_{D} is given by $g \in H^{\frac{1}{2}}(\Gamma_{\mathrm{D}}; \mathbb{R}^3)$, and the surface outer force on Γ_{N} is given by $f^{\mathrm{N}} \in L^2(\Gamma_{\mathrm{D}}; \mathbb{R}^3)$, as prescribed boundary values.

Under suitable regularity conditions, it is well-known that a unique solution to the problem (6) is given by the unique minimizer of a total elastic energy

$$E_1(u) := \int_{\Omega} \{w[u] - f \cdot u\}\,\mathrm{d}x - \int_{\Gamma_{\mathrm{N}}} f^{\mathrm{N}} \cdot u\,\mathrm{d}s, \tag{7}$$

with the boundary condition $u = g$ on Γ_{D}, where $w[u]$ is the strain energy density defined as $w[u] := \frac{1}{2}\sigma[u] : e[u]$.

In case that the elastic material is isotropic and homogeneous, the elasticity tensor has the form

$$c_{ijkl} = \lambda\delta_{ij}\delta_{kl} + \mu(\delta_{ik}\delta_{jl} + \delta_{il}\delta_{jk}),$$

where $\lambda \geq 0$ and $\mu > 0$ are called the Lamé constants. Since

$$c_{ijkl}\xi_{ij}\xi_{kl} = \lambda(\operatorname{tr}\xi)^2 + 2\mu|\xi|^2, \qquad \xi \in \mathbb{R}^{3\times 3}_{\mathrm{sym}},$$

the condition (5) is satisfied with $c_* = 2\mu$. The stress tensor and the strain energy density become

$$\sigma[u] = \lambda(\operatorname{div} u)I + 2\mu e[u], \qquad w[u] = \frac{1}{2}\lambda(\operatorname{div} u)^2 + \mu|e[u]|^2,$$

where I denotes the unit tensor. The equilibrium equations of force are often called the Navier or Navier-Cauchy equations and take the following form:

$$-\mu\Delta u - (\lambda + \mu)\nabla(\operatorname{div} u) = f, \qquad \text{in } \Omega. \tag{8}$$

Let us now consider the case that Ω contains a dislocation curve $\Gamma(t)$. We assume that $\Gamma(t)$ is a closed plane curve without self-intersections in a fixed crystallographic plane $\tilde{S} \subset \mathbb{R}^3$ and define $S := \tilde{S} \cap \Omega$, which is called the slip plane. We suppose that S is connected and open in $\tilde{S}$. The Burgers vector of $\Gamma(t)$ is denoted by $b \in \mathbb{R}^3 \setminus \{\mathbf{0}\}$ which is a fixed vector tangential to S. This is a so-called mixed dislocation which contains both edge and screw dislocations at the parts of $\Gamma(t)$ where $b \perp \Gamma(t)$ and $b \parallel \Gamma(t)$, respectively.

Choosing a suitable orthogonal coordinate system, without loss of generality, we suppose that $\tilde{S} = \{x = (x_1, x_2, x_3)^{\mathrm{T}};\ x_3 = 0\}$, $b = (b_1, 0, 0)^{\mathrm{T}}$, and $b_1 > 0$. A typical example of Ω and S is shown in Figs. 1 and 2, where

$$\Omega = S_0 \times (-H, H), \qquad \Gamma_{\mathrm{D}} = S_0 \times \{-H, H\}, \qquad \Gamma_{\mathrm{N}} = \partial S_0 \times [-H, H].$$

We often identify the slip plane $S = S_0 \times \{0\}$ with $S_0 \subset \mathbb{R}^2$, if no confusion occurs. The coordinate in $S \cong S_0 \subset \mathbb{R}^2$ is denoted by $x' = (x_1, x_2)^{\mathrm{T}}$. We consider $\Gamma(t)$ and S as subsets of $\mathbb{R}^2$, and a two dimensional domain enclosed by $\Gamma(t)$ in S is denoted by $D(t) \subset S \subset \mathbb{R}^2$.

We define $\Omega^{\pm} := \{(x_1, x_2, x_3)^{\mathrm{T}} \in \Omega;\ \pm x_3 > 0\}$ and denote the outward unit normal vectors on $\partial\Omega^{\pm}$ by $\nu^{\pm}$, respectively. It is considered that the displacement field u is discontinuous across the slip plane S. The traces of u to S from Ω^+ and Ω^- are denoted by u^+ and u^-, respectively, and normal tractions $\sigma[u]\nu$ from Ω^+ and Ω^- on S are denoted by $\sigma^+\nu$ and $\sigma^-\nu$, respectively, where we define $\nu = (0, 0, 1)^{\mathrm{T}}$ on S. We also denote the outward unit normal vector on $\partial\Omega$ by ν. We remark that $\nu = \nu^{\pm}$ on $\partial\Omega \cap \partial\Omega^{\pm}$ and that $\nu = \mp\nu^{\pm}$ on S. The gaps of the displacement and the traction across the slip plane S are denoted by $\langle u \rangle_S := u^+ - u^-$ and $\langle \sigma \rangle_S \nu := \sigma^+\nu - \sigma^-\nu$, respectively.

For a fixed time t, it is naturally expected that the displacement field $u(x)$ in Ω satisfies the following boundary value problem in a naive setting:

$$\begin{cases} -\partial_i \sigma_{ij}[u] = f_j(x), \quad j = 1,2,3, & x \in \Omega^+ \cup \Omega^-, \\ u = g(x), & x \in \Gamma_{\mathrm{D}}, \\ \sigma[u]\nu = f^{\mathrm{N}}(x), & x \in \Gamma_{\mathrm{N}}, \\ \langle u \rangle_S = \begin{cases} b, & \text{on } D(t), \\ 0, & \text{on } S \setminus D(t), \end{cases} \\ \langle \sigma[u] \rangle_S \nu = 0, \quad \text{on } S. \end{cases} \tag{9}$$

This problem, however, has no finite energy solution as seen below. Let us suppose that u is a finite energy solution, i.e., $u|_{\Omega^\pm} \in H^1(\Omega^\pm; \mathbb{R}^3)$. Then its traces $u^\pm$ and $\langle u \rangle_S$ should belong to $H^{1/2}(S; \mathbb{R}^3)$ but it contradicts the fourth condition of (9). In the next section, we consider a weak formulation of a jump problem in general form and study the condition that its solution has a finite energy.

Weak Formulations for Jump Problems

We study a weak formulation for a linear elasticity problem with a jump condition across an interface like (9) in a general setting.

Under the same condition of (9), for a $\mathbb{R}^3$-valued function Ψ on S, we consider the following problem

$$\begin{cases} -\partial_j \sigma_{ij}[u] = f_i(x), \quad i = 1,2,3, & x \in \Omega^+ \cup \Omega^-, \\ u = g(x), & x \in \Gamma_{\mathrm{D}}, \\ \sigma[u]\nu = f^{\mathrm{N}}(x), & x \in \Gamma_{\mathrm{N}}, \\ \langle u \rangle_S = \Psi, & \text{on } S, \\ \langle \sigma[u] \rangle_S \nu = 0, & \text{on } S. \end{cases} \tag{10}$$

We suppose that $\Psi \in H^{\frac{1}{2}}(S; \mathbb{R}^3)$ and $g \in H^{\frac{1}{2}}(\Gamma_{\mathrm{D}}; \mathbb{R}^3)$. For the problem (10), we define function spaces:

$$\begin{aligned} X &:= \{ w \in L^2(\Omega; \mathbb{R}^3);\ w|_{\Omega^\pm} \in H^1(\Omega^\pm; \mathbb{R}^3) \}, \\ X(\Psi) &:= \{ w \in X;\ \langle w \rangle_S = \Psi \}, \\ V(g, \Psi) &:= \{ w \in X(\Psi);\ w = g \text{ on } \Gamma_{\mathrm{D}} \}, \\ V &:= V(0,0), \end{aligned}$$

where $\langle w\rangle_S = w^+ - w^-$ on S for $w \in X$. The space X is a Hilbert space with the following norm and a corresponding inner product:

$$\|w\|_X := \left(\|w\|^2_{H^1(\Omega^+;\mathbb{R}^3)} + \|w\|^2_{H^1(\Omega^-;\mathbb{R}^3)}\right)^{\frac{1}{2}}.$$

We can identify $X(0) = H^1(\Omega;\mathbb{R}^3)$. We remark that X is a space of finite energy displacements with a gap across S. In the following lines, we suppose that $V(0,\Psi) \neq \emptyset$.

If $u|_{\Omega^\pm} \in H^2(\Omega^\pm;\mathbb{R}^3)$, and u satisfies the equations of (10), u is called a strong solution to (10), where the boundary conditions are considered in the sense of the trace operator. We also define a weak solution as follows.

Problem 2.1 *Find $u \in V(g,\Psi)$ such that*

$$\int_{\Omega\setminus S} \sigma[u] : e[w]\,\mathrm{d}x - \int_\Omega f\cdot w\,\mathrm{d}x - \int_{\Gamma_\mathrm{N}} f^\mathrm{N}\cdot w\,\mathrm{d}s = 0, \quad \textit{for all} \quad w \in V. \tag{11}$$

A solution of Problem 2.1 is called a weak solution to (10). In particular, it is not difficult to show that a strong solution to (10) is a weak solution, by a standard computation with integration by parts. More precisely, we have the following theorem.

Theorem 2.2 *Suppose that $f \in L^2(\Omega;\mathbb{R}^3)$, $f^\mathrm{N} \in L^2(\Gamma_\mathrm{N},\mathbb{R}^3)$, $g \in H^{\frac{1}{2}}(\Gamma_\mathrm{D};\mathbb{R}^3)$ and $\Psi \in H^{\frac{1}{2}}(S;\mathbb{R}^3)$. Then, u is a strong solution to* (10) *if and only if u is a weak solution and $u|_{\Omega^+}$ and $u|_{\Omega^-}$ belong to $H^2(\Omega^+;\mathbb{R}^3)$ and $H^2(\Omega^-;\mathbb{R}^3)$, respectively.*

The unique existence of the weak solution is guaranteed as follows.

Theorem 2.3 *Under the same conditions of Theorem 2.2, there exists a unique solution u to Problem 2.1 and u is given as a unique minimizer of $E_1(u)$ in $V(g,\Psi)$.*

We omit the proofs of these theorems here and postpone them until our forthcoming paper. Here we just admit that the weak solution uniquely exists and let our argument proceed.

On the other hand, if Ψ does not belong to $H^{1/2}(S;\mathbb{R}^3)$ as in the dislocation model (9), there is no finite energy solution since Ψ cannot be expressed by $\langle u\rangle_S$ for any $u \in X$. According to the model (9), the displacement field u has to have a singularity along the dislocation curve with infinite elastic energy, and of course this is not a realistic solution. In a microscopic description of a crystal lattice, it is considered that this singularity is somehow regularized by the existence of a minimum length corresponding to the height of an atom. In the next section, we introduce a regularization in terms of phase field approach.

Regularization of Energy

We consider a regularization for the curvature flow model (1) by means of an Allen-Cahn-type bistable potential energy. Let φ be a phase field which is a smooth scalar-valued function defined on $S \times [0, T)$ and the value is approximately 1 in $D(t)$ and 0 in $S \setminus \overline{D(t)}$. For positive parameters $\varepsilon > 0$ and $\beta > 0$, we define an interface energy as

$$E_0(\varphi) := \int_S \Big(\frac{\varepsilon}{2}|\nabla'\varphi|^2 + \beta W(\varphi)\Big)\,\mathrm{d}x', \tag{12}$$

where $W(s) = s^2(1-s)^2$. As its gradient flow in $L^2(S)$, we have the following Allen-Cahn equation

$$\alpha\frac{\partial\varphi}{\partial t} = \varepsilon\Delta'\varphi - \beta W'(\varphi), \qquad \text{in } S \times (0, T), \tag{13}$$

where ∇' and Δ' represent the gradient and the Laplacian with respect to $x' = (x_1, x_2)^{\mathrm{T}}$. It is known that the interface energy (12) and the Allen-Cahn equation correspond to the length of the curve $\Gamma(t)$ and a motion by line tension, respectively, under a suitable scaling [6].

Using this regularized interface energy $E_0(\varphi)$ together with the elastic energy $E_1(u)$ defined by (7), we consider the following total energy:

$$E(\varphi) := E_0(\varphi) + E_1(u_\varphi), \qquad \varphi \in H^1(S),$$

where u_φ is a unique solution of (11) with $u_\varphi \in V(g, b\varphi)$.

Let us derive an L^2 gradient flow of the energy $E(\varphi)$. For a smooth scalar function ψ defined on S, we consider a first variation:

$$\frac{d}{dr}E(\varphi + r\psi)|_{r=0} = \frac{d}{dr}E_0(\varphi + r\psi)|_{r=0} + \frac{d}{dr}E_1(u_{\varphi+r\psi})|_{r=0}. \tag{14}$$

The first term is formally given as

$$\frac{d}{dr}E_0(\varphi + r\psi)|_{r=0} = \int_S \big(-\varepsilon\Delta'\varphi + \beta W'(\varphi)\big)\psi\,\mathrm{d}x' + \int_{\partial S}\frac{\partial\varphi}{\partial\nu'}\psi\,\mathrm{d}s',$$

where $\nu' \in \mathbb{R}^2$ denotes the outward unit normal vector on $\partial S \cong \partial S_0$. For the second term of (14), it is easy to show that $u_{\varphi+r\psi} = u_\varphi + ru_* \in V(g, b(\varphi + r\psi))$ where u_* is a unique solution of the following weak form:

$$u_* \in V(0, b\psi), \qquad \int_{\Omega\setminus S}\sigma[u_*] : e[w]\,\mathrm{d}x = 0, \quad \text{for all} \quad w \in V. \tag{15}$$

Then, under suitable regularity conditions, we obtain

$$
\begin{aligned}
\frac{d}{dr}E_1(u_{\varphi+r\psi})|_{r=0} &= \frac{d}{dr}E_1(u_\varphi + ru_*)|_{r=0} \\
&= \int_{\Omega^+\cup\Omega^-} \{\sigma[u_\varphi] : e[u_*] - f\cdot u_*\}\,\mathrm{d}x - \int_{\Gamma_\mathrm{N}} f^\mathrm{N}\cdot u_*\,\mathrm{d}s \\
&= \int_{\partial\Omega^+} (\sigma[u_\varphi]\nu^+)\cdot u_*\,\mathrm{d}s + \int_{\partial\Omega^-} (\sigma[u_\varphi]\nu^-)\cdot u_*\,\mathrm{d}s \\
&\quad - \int_{\Omega^+\cup\Omega^-} \{\operatorname{div}\sigma[u_\varphi] + f\}\cdot u_*\,\mathrm{d}x - \int_{\Gamma_\mathrm{N}} f^\mathrm{N}\cdot u_*\,\mathrm{d}s \\
&= \int_S \{(-\sigma^+[u_\varphi]\nu)\cdot u_*^+ + (\sigma^-[u_\varphi]\nu)\cdot u_*^-\}\,\mathrm{d}x' \\
&\quad + \int_{\Gamma_\mathrm{N}} (\sigma[u_\varphi]\nu - f^\mathrm{N})\cdot u_*\,\mathrm{d}s \\
&= -\int_S (\sigma[u_\varphi]\nu)\cdot\langle u_*\rangle_S\,\mathrm{d}x' \\
&= -\int_S (\sigma[u_\varphi]\nu)\cdot b\psi\,\mathrm{d}x'.
\end{aligned}
$$

Hence, using the homogeneous Neumann boundary condition $\frac{\partial\varphi}{\partial\nu'} = 0$ on ∂S, we obtain

$$
\frac{d}{dr}E(\varphi + r\psi)|_{r=0} = \int_S \big(-\varepsilon\Delta'\varphi + \beta W'(\varphi) - (\sigma[u_\varphi]\nu)\cdot b\big)\psi\,\mathrm{d}x'.
$$

Similarly to the Allen-Cahn equation (13), we derive a gradient flow of the energy $E(\varphi)$. Therefore, we propose the following phase field model for dislocation dynamics

$$
\begin{cases}
\alpha\dfrac{\partial\varphi}{\partial t} = \varepsilon\Delta'\varphi - \beta W'(\varphi) + (\sigma[u_\varphi]\nu)\cdot b, & \text{in } S\times(0,T), \\
\dfrac{\partial\varphi}{\partial\nu'} = 0, & \text{on } \partial S\times(0,T), \\
\varphi(\cdot,0) = \varphi_0, & \text{on } S,
\end{cases}
\tag{16}
$$

where $\varphi_0(x')$ is a suitable initial value for φ. The additional force term $(\sigma[u_\varphi]\nu)\cdot b$ appearing above is nothing but the Peach-Koehler force (3) acting on the dislocation line. The model (16) naturally contains the notion of the dislocation core and the

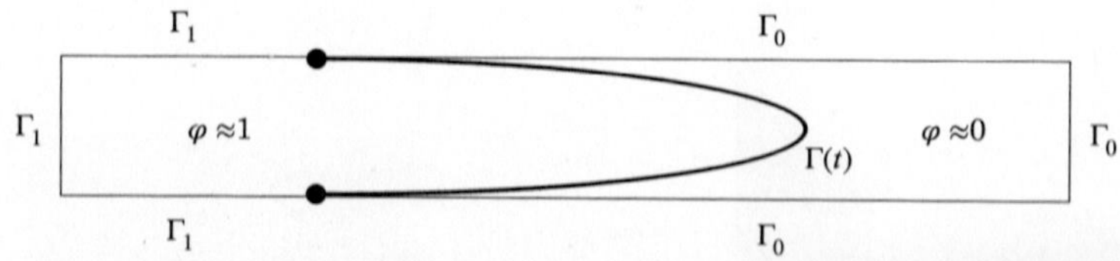

Fig. 3 Another possible configuration of the dislocation curve with fixed end points in our model is shown. The Dirichlet boundary condition for φ is imposed as $\varphi|_{\Gamma_0} \approx 0$ and $\varphi|_{\Gamma_1} \approx 1$

Peach-Koehler force. The gradient flow structure

$$\frac{d}{dt}E(\varphi(\cdot,t)) = -\alpha \int_S \left| \frac{\partial \varphi}{\partial t}(x',t) \right|^2 dx' \leq 0, \tag{17}$$

behind the dynamics of dislocation curve has been revealed in the derivation of the model as above.

Remark In this paper, we treat only the case where the dislocation is a closed plane curve without any end point in the slip plane. On the other hand, it is often observed in real materials that two end points are fixed at some defect of the crystal structure. It is also known that the dislocation curve sometimes can change its slip plane. For example, some numerical simulations of dislocations with end points which change their slip planes are shown in [10]. The model presented in this paper can also be applied in the case where the end points are fixed on the boundary of the slip plane, if we impose the Dirichlet boundary condition of the phase field variable φ as illustrated in Fig. 3. However, the treatment of the change of the slip plane as shown in [10] in our model seems challenging but difficult at present.

3 A Simplified 2D-1D Model

In this section, we derive a 2D-1D coupled phase field model for dynamics of a straight screw dislocation line in the same manner as the 3D-2D model of the previous section.

As shown in Fig. 4, we consider a rectangular parallelepiped elastic body and a slip plane in x_1x_2 plane. A straight screw dislocation with the Burgers vector $(b, 0, 0)^{\mathrm{T}}$ is assumed to be moving in the slip plane. In the following lines, we denote the coordinate (x_2, x_3) also by (x, y) for simplicity. For the displacement field, we suppose the so-called anti-plane displacement and an odd symmetry with respect to the slip plane $y = 0$, i.e., the displacement has the form $(u(x, y), 0, 0)^{\mathrm{T}}$ and $u(x, -y) = -u(x, y)$. Then the equilibrium equation (8) becomes $-\mu \Delta u = 0$, where no body force is supposed.

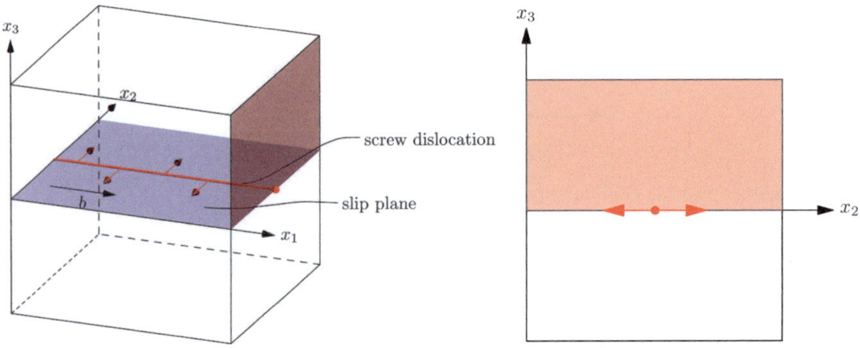

Fig. 4 A straight screw dislocation in a slip plane and its two dimensional projection

In the following sections, we set a rectangular two-dimensional domain $\Omega := (-L, L) \times (0, H)$ in xy plane. We define the bottom boundary $S := (-L, L) \times \{0\}$, the lateral boundary $\Gamma_0 := \{-L, L\} \times [0, H]$, and top boundary $\Gamma_1 := (-L, L) \times \{H\}$. Then $\partial\Omega = S \cup \Gamma_0 \cup \Gamma_1$ holds. The Laplacian with respect to (x, y) is denoted by $\Delta = \frac{\partial^2}{\partial x^2} + \frac{\partial^2}{\partial y^2}$.

Let $u(x, y, t) \in \mathbb{R}$ be an anti-plane displacement at time t and position $(x, y) \in \Omega$. Similarly to section "Regularization of Energy", we introduce a phase field variable $\varphi \in H^1(-L, L)$ and define the following total energy. We suppose that an anti-plane boundary traction $g(x)$ is given on the top boundary Γ_1. We set $F(s) = s^2(1-s)^2$ and define

$$
\begin{aligned}
E(\varphi) &:= E_0(\varphi) + E_1(u_\varphi),\\
E_0(\varphi) &:= \int_{-L}^{L} \left(\frac{\varepsilon}{2}|\varphi_x|^2 + \beta F(\varphi)\right) \mathrm{d}x,\\
E_1(u) &:= \mu \iint_{\Omega} |\nabla u|^2 \,\mathrm{d}x\,\mathrm{d}y - 2\int_{\Gamma_1} g u \,\mathrm{d}x,
\end{aligned}
$$

where $u_\varphi \in H^1(\Omega)$ is given as the unique weak solution to

$$
\begin{cases}
\Delta u = 0, & \text{in } \Omega,\\
u_x = 0, & \text{on } \Gamma_0,\\
\mu u_y = g(x), & \text{on } \Gamma_1,\\
u = \dfrac{b}{2}\varphi(x), & \text{on } S.
\end{cases}
\tag{18}
$$

Let us derive an L^2-gradient flow of the energy $E(\varphi)$. Under the Neumann boundary condition $\varphi_x(-L) = \varphi_x(L) = 0$, for a smooth function ψ defined on

S, we consider a first variation

$$\frac{d}{dr}E(\varphi + r\psi)|_{r=0} = \frac{d}{dr}E_0(\varphi + r\psi)|_{r=0} + \frac{d}{dr}E_1(u_{\varphi+r\psi})|_{r=0}. \tag{19}$$

The first term is formally given as

$$\frac{d}{dr}E_0(\varphi + r\psi)|_{r=0} = \int_{-L}^{L} (-\varepsilon\varphi_{xx} + \beta F'(\varphi))\, \psi \, \mathrm{d}x.$$

It is easy to show that $u_{\varphi+r\psi} = u_\varphi + \frac{1}{2}rbw \in V(g, \varphi + r\psi)$, where $w \in H^1(\Omega)$ is the unique weak solution of

$$\begin{cases} \Delta u = 0, & \text{in } \Omega, \\ u_x = 0, & \text{on } \Gamma_0, \\ u_y = 0, & \text{on } \Gamma_1, \\ u = \psi(x), & \text{on } S. \end{cases}$$

Then the second term of (19) becomes

$$\begin{aligned} \frac{d}{dr}E_1(u_{\varphi+r\psi})|_{r=0} &= \frac{d}{dr}E_1(u_\varphi + \frac{1}{2}rbw)|_{r=0} \\ &= \mu b \iint_\Omega \nabla u_\varphi \cdot \nabla w \, \mathrm{d}x\, \mathrm{d}y - b\int_{\Gamma_1} gw \, \mathrm{d}x \\ &= \mu b \int_{\partial\Omega} \frac{\partial u_\varphi}{\partial \nu} w \, \mathrm{d}s - \mu b \iint_\Omega (\Delta u_\varphi) w \, \mathrm{d}x\, \mathrm{d}y - b\int_{\Gamma_1} gw \, \mathrm{d}x \\ &= -\mu b \int_S \frac{\partial u_\varphi}{\partial y} \psi \, \mathrm{d}x. \end{aligned}$$

Hence, we obtain

$$\frac{d}{dr}E(\varphi + s\psi)|_{r=0} = \int_{-L}^{L} \left(-\varepsilon\varphi_{xx} + \beta F'(\varphi) - \mu b \frac{\partial u_\varphi}{\partial y}(x, 0)\right) \psi \, \mathrm{d}x.$$

Similarly to the Allen-Cahn equation (13), we derive a gradient flow of the energy $E(\varphi)$ as

$$\alpha\varphi_t = \varepsilon\varphi_{xx} - \beta F'(\varphi) + b\mu \frac{\partial u_\varphi}{\partial y}(x, 0), \qquad x \in (-L, L),\ t > 0. \tag{20}$$

The additional force term $b\mu\frac{\partial u_\varphi}{\partial y}(x,0)$ is formally corresponding to the Peach-Koehler force (3) acting on the screw dislocation. We substitute the relation

$$\varphi(x,t) = \frac{2}{b}u_\varphi(x,0,t), \qquad x \in [-L,L],\ t \geq 0,$$

into (20) and set

$$W(s) := \frac{4\beta}{b^2}s^2\left(\frac{b}{2}-s\right)^2, \qquad \gamma := \frac{\mu b^2}{2}.$$

The boundary load $g(x)$ is also assumed to depend on time t. Then, we obtain the following 2D-1D dislocation model:

$$\begin{cases} \Delta u = 0, & (x,y) \in \Omega,\ t \in [0,T], \\ u_x = 0, & (x,y) \in \Gamma_0,\ t \in [0,T], \\ u_y = g(x,t), & (x,y) \in \Gamma_1,\ t \in [0,T], \\ \alpha u_t = \varepsilon u_{xx} - W'(u) + \gamma u_y, & (x,y) \in S,\ t \in (0,T], \\ u = u_0(x), & (x,y) \in S,\ t = 0. \end{cases} \tag{21}$$

4 Numerical Results

In this section, we give some numerical examples for an approximation of (21).

We discretize (21) by employing standard finite differences to approximate spatial derivatives. A rectilinear grid is obtained by dividing Ω into $N_x \times N_y$ rectangular cells and a solution is sought at the grid nodes. This spatial discretization results in a non-linear system of ODEs in time that we solve by means of a fully implicit, variable-step solver [5].

The initial condition in both examples below is a shifted and scaled Heaviside function $u_0(x) = bH(x - x_0)$ where x_0 is the initial position of the step. Other parameter values used in the numerical examples are summarized in Table 1.

In both figures below the horizontal axis corresponds to the spatial x variable while the vertical axis corresponds to the time. We plot the graph of u at S at 50 time levels distributed uniformly in $[0,T]$, the initial condition u_0 being the lowest graph with the time increasing upwards. The graph of u in the whole of Ω is not shown.

Table 1 Parameter values used in numerical examples for the model (21)

T	L	H	N_x	N_y	b	ε	β	μ	α
4	2	2	256	128	0.06	0.04	10	10	0.01

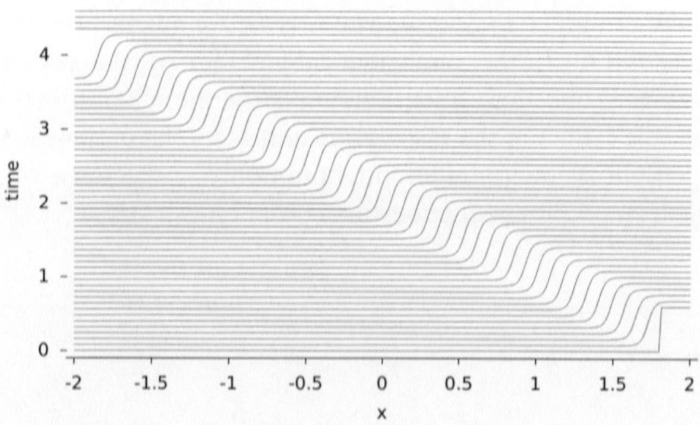

Fig. 5 Numerical example of a screw dislocation moving under a constant loading

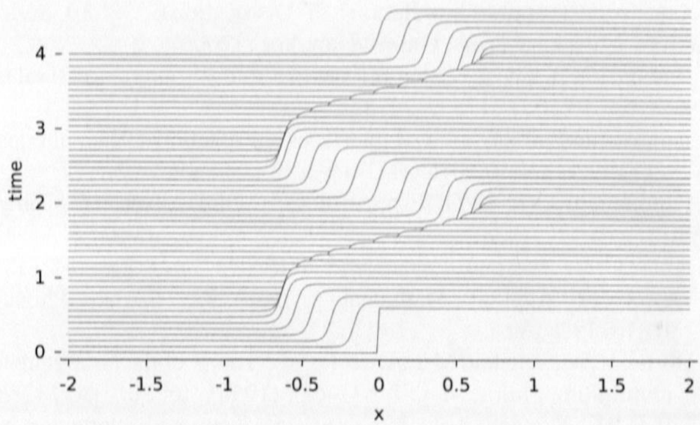

Fig. 6 Numerical example of a screw dislocation moving under a periodic loading

In Fig. 5 we present a numerical example of a screw dislocation that moves in a stress field caused by a constant anti-plane traction imposed on the top boundary Γ_1. We set $g \equiv 0.5$ and place the dislocation at the initial location $x_0 = 1.8$. The dislocation moves to the left at an almost constant velocity until it reaches the crystal surface at $x = -2$ where it annihilates.

In Fig. 6 we consider the behaviour of a screw dislocation under a periodic loading. We place the dislocation at $x_0 = 0$ and set $g(x, t) = \cos(0.5t)$. The oscillating stress field causes the dislocation to move periodically around its initial location.

5 Summary and Conclusions

We proposed a 3D-2D phase field model (16) for dynamics of a mixed dislocation curve and a 2D-1D model (21) for a straight screw dislocation. They are both derived as gradient flows of regularized total energies and naturally include the Peach-Koehler force and the notion of the dislocation core. Some numerical examples of the 2D-1D model are given in Sect. 4. The revealed gradient flow structure (17) is expected to be useful for further mathematical and numerical analysis.

One of interesting questions about this model is the relation between the layer width of φ (i.e., the radius of the dislocation core) and the parameters ε, β and $|b|$. This is still open but seems to be important not only in the sense of modeling but also in numerical simulations for choosing a proper mesh size.

References

1. Alvarez, O., Carlini, E., Hoch, P., Le Bouar, Y., Monneau, R.: Dislocation dynamics described by non-local Hamilton-Jacobi equations. Mater. Sci. Eng. A **400–401**, 162–165 (2005). doi:10.1016/j.msea.2005.01.062
2. Bulatov, V., Cai, W.: Computer Simulations of Dislocations. Oxford Series on Materials Modelling. Oxford University Press, Oxford/New York (2006)
3. Carter, W.C., Taylor, J.E., Cahn, J.W.: Variational methods for microstructural evolutions. JOM **49**, 30–26 (1997). doi:10.1007/s11837-997-0027-2
4. Chen, X.: Generation and propagation of interfaces for reaction-diffusion equations. J. Differ. Equ. **96**, 116–141 (1992). doi:10.1016/0022-0396(92)90146-E
5. Cohen, S.D., Hindmarsh, A.C.: CVODE, a stiff/nonstiff ODE solver in C. Comput. Phys. **10**(2), 138–143 (1996)
6. Fife, P.C.: Dynamics of Internal Layers and Diffusive Interfaces. CBMS-NSF Regional Conference Series in Applied Mathematics, vol. 53. SIAM, Philadelphia (1988). doi:10.1137/1.9781611970180
7. Hirsch, P.B., Horne, R.W., Whelan, M.J.: Direct observations of the arrangement and motion of dislocations in aluminium. Philos. Mag. **1**, 677–684 (1956). doi:10.1080/14786435608244003
8. Minárik, V., Beneš, M., Kratochvíl, J.: Simulation of dynamical interaction between dislocations and dipolar loops. J. Appl. Phys. **107**, 061802 (2010). doi:10.1063/1.3340518
9. Nabarro, F.R.N.: Theory of Crystal Dislocations. Clarendon Press, Oxford (1967)
10. Pauš, P., Beneš, M., Kratochvíl, J.: Simulation of dislocation annihilation by cross-slip. Acta Physica Polonica A **122**(3), 509–5011 (2012)

Persistence of Common Topological Structures by Commutative Triple Ladder Quiver

Emerson G. Escolar and Yasuaki Hiraoka

Abstract This is a summary paper of Escolar and Hiraoka (Persistence modules on commutative ladders of finite type. Discrete Comput Geom 55, 100–157 (2016)) which presents an extension of persistence modules as representations on quivers with nontrivial relations. In particular, the mathematical and algorithmic results in that paper enable us to detect robust and common topological structures of two geometric objects. In this paper, we only deal with a special type of persistence modules defined on the so-called commutative triple ladder for the sake of simplicity. We aim to explain the essence of Auslander-Reiten theory in connection with persistence modules.

Keywords Persistence modules • Commutative triple ladder • Representation theory • Topological data analysis

1 Introduction

This article summarizes the paper [7] that presents a novel method to detect robust and common topological features of two geometric objects. The main ideas are to extend the notion of persistent homology [6, 13] to representations on a special type of quiver and to study its indecomposable decompositions. The result can be regarded as a partial extension towards multi-dimensional persistent homology (e.g., see [4]) and enables one to study two-parameter persistence.

Let X_r and Y_r be two topological spaces parametrized by $r \in \mathbb{N}$. Suppose that, for parameters $r < s$, we have continuous maps $f : X_r \to X_s$ and $g : Y_r \to Y_s$. By taking homology $f_* : H_*(X_r) \to H_*(X_s)$ and $g_* : H_*(Y_r) \to H_*(Y_s)$ with field coefficient K, we can study persistent topological features in the parameter interval

E.G. Escolar (✉) • Y. Hiraoka
WPI – Advanced Institute for Materials Research (WPI-AIMR), Tohoku University, 2-1-1, Katahira, Aoba-ku, Sendai, 980-8577, Japan
e-mail: escolar.emerson.gaw.d6@tohoku.ac.jp; hiraoka@wpi-aimr.tohoku.ac.jp

Y. Nishiura, M. Kotani (eds.), *Mathematical Challenges in a New Phase of Materials Science*, Springer Proceedings in Mathematics & Statistics 166,
DOI 10.1007/978-4-431-56104-0_4

$[r, s]$ for both X and Y. This is the simplest setting of persistent homology [6, 13], and is characterized by a unique decomposition

$$(f_* : H_*(X_r) \to H_*(X_s)) \quad \simeq \quad I(r, r)^{n_r} \oplus I(r, s)^{n_{rs}} \oplus I(s, s)^{n_s},$$

where $n_r, n_{rs}, n_s \in \mathbb{N}_0 = \{0, 1, 2, \dots\}$, and $I(r, r), I(r, s)$, and $I(s, s)$ are indecomposable modules defined by

$$I(r, r) : K \to 0, \quad I(r, s) : K \to K, \quad I(s, s) : 0 \to K.$$

Similarly, we obtain a decomposition for $g_* : H_*(Y_r) \to H_*(Y_s)$. These decompositions characterize "persistence" in the sense that topological features specified by the summand $I(r, s)$ are regarded as robust in the parameter interval $[r, s]$, and those specified by the summands $I(r, r)$ and $I(s, s)$ are observed only at each parameter value. Our interest in this paper is to compare the persistent topological features of $f : X_r \to X_s$ and $g : Y_r \to Y_s$.

For this purpose, it is natural to map X_r and Y_r into another topological space Z_r as

$$X_r \to Z_r \leftarrow Y_r,$$

and study common topological features via Z_r. One of the standard choices for these maps is the inclusion $X_r \hookrightarrow X_r \cup Y_r \hookleftarrow Y_r$. Then, in order to study robust and common topological features, a commutative diagram of topological spaces

$$\begin{array}{ccccc} X_s & \longrightarrow & Z_s & \longleftarrow & Y_s \\ \uparrow & & \uparrow & & \uparrow \\ X_r & \longrightarrow & Z_r & \longleftarrow & Y_r \end{array} \tag{1}$$

provides an appropriate geometric setting. It should be noted that the individual robust topological features are measured in the vertical direction and the common topological features are measured in the horizontal direction. Hence, on the homological level, our problem is to consider a commutative diagram

$$\begin{array}{ccccc} H_*(X_s) & \longrightarrow & H_*(Z_s) & \longleftarrow & H_*(Y_s) \\ \uparrow & & \uparrow & & \uparrow \\ H_*(X_r) & \longrightarrow & H_*(Z_r) & \longleftarrow & H_*(Y_r) \end{array} \tag{2}$$

of K-vector spaces and linear maps between them.

The algebraic object (2) can be regarded as a representation of the quiver

$$\begin{array}{ccccc} \circ & \longrightarrow & \circ & \longleftarrow & \circ \\ \uparrow & \circlearrowleft & \uparrow & \circlearrowleft & \uparrow \\ \circ & \longrightarrow & \circ & \longleftarrow & \circ \end{array} \tag{3}$$

with commutative relations. We call this quiver the commutative triple ladder in this paper. Recall that a representation V of this quiver is given by a commutative diagram

$$\begin{array}{ccccc} V_4 & \xrightarrow{f_{45}} & V_5 & \xleftarrow{f_{65}} & V_6 \\ {\scriptstyle f_{14}}\uparrow & & \uparrow{\scriptstyle f_{25}} & & \uparrow{\scriptstyle f_{36}} \\ V_1 & \xrightarrow[f_{12}]{} & V_2 & \xleftarrow[f_{32}]{} & V_3 \end{array} \tag{4}$$

of finite dimensional K-vector spaces and linear maps between them. In view of the correspondence between representations and modules, we call (4) a persistence module [3, 4, 7, 13] on the commutative ladder (3).

Under this setting, the paper [7] clarifies the following:

(i) The number of isomorphism classes of indecomposable persistence modules on (3) is finite. This implies that persistence modules on (3) can be classified by complete discrete invariants.

(ii) The Auslander-Reiten quiver of (3), which lists up all the isomorphism classes of indecomposable persistence modules and irreducible morphisms among them, is explicitly derived. Moreover, the notion of persistence diagrams is generalized to functions on the vertex set of the Auslander-Reiten quiver. In particular, the multiplicity of the persistence module

$$\begin{array}{ccccc} K & \longrightarrow & K & \longleftarrow & K \\ \uparrow & & \uparrow & & \uparrow \\ K & \longrightarrow & K & \longleftarrow & K, \end{array}$$

where all the maps are identity maps, characterizes the robust and common topological features between $f : X_r \to X_s$ and $g : Y_r \to Y_s$.

(iii) An algorithm for computing indecomposable decompositions is presented by using the Auslander-Reiten quiver.

(iv) Numerical examples to detect robust and common topological features are shown.

We note that the quiver (3) is not one of A_n, D_n and E_6, E_7, E_8 types studied in Gabriel's theorem [8]. However, the commutativity imposed in (3) enables us to derive the finiteness property of the representation type as mentioned above.

The research in [7] is motivated both by a recent theoretical result and by applications of topological data analysis. On the theoretical side, it is inspired by the paper [3], where Carlsson and de Silva formulate persistent homology as representations of an A_n quiver and generalize to zigzag persistence. By this connection to the representation theory, the algebraic aspects of persistent homology have been understood further. The paper [7] deals with a generalization to representations of associative algebras.

On the other hand, from the application side, this work is motivated by the analysis of protein [9] and amorphous structures [10, 11] using persistent homology. For example, in the paper [9], the authors study a topological characterization of protein compressibility by using persistent homology computed on α-complex models of proteins. One of the key observations obtained in that work is that there exist two distinct radius parameters $r < s$ for the atoms which characterize protein compressibility. Hence, once we develop a tool which gives us more detailed information of robust and common topological features on these restricted two parameter values, we expect to obtain further understanding of the relationship between protein compressibility and their geometric structures. The result in the paper [7] provides a tool for that purpose.

This paper is organized as follows. In Sect. 2, we introduce necessary background from persistent homology and representation theory. Then, the main theorem (Theorem 3) is presented in this section, together with a summary of the technique used for its proof. In Sect. 3, we sketch an algorithm for computing indecomposable decompositions. Section 4 presents a numerical example.

2 Commutative Triple Ladder Persistence

Persistence and Representation Theories

In this section, we recall fundamental notions of persistent homology and representation theory. For details of the representation theory of associative algebras, we refer to [1, 2, 12]. We here note that the words "module" and "representation" are used interchangeably by the categorical equivalence between them.

We define a persistence module on (3) as its representation (see (4) and the exposition below it). We assume that vector spaces in the representation are finite dimensional. For persistence modules V and V' on (3), a morphism $\varphi : V \to V'$ is defined by a set of linear maps $\varphi_i : V_i \to V'_i$ satisfying $f'_{ij} \circ \varphi_i = \varphi_j \circ f_{ij}$ for all edges $\overrightarrow{ij}$, where f'_{ij} are the linear maps in V'. A morphism φ is called an isomorphism if there exists a morphism $\psi : V' \to V$ such that $\psi \circ \varphi = 1$ and $\varphi \circ \psi = 1$. If there exists an isomorphism between V and V', they are called isomorphic and denoted by $V \simeq V'$.

A submodule W of V is defined by a set of vector subspaces $W_i \subset V_i$ such that $f_{ij}(W_i) \subset W_j$. These subspaces themselves define a module on (3) by restrictions of linear maps. A submodule W is called a direct summand if there exists a submodule W' of V such that $V_i \simeq W_i \oplus W_i'$ at each vertex i. In this case, we write $V \simeq W \oplus W'$ and call this a direct sum decomposition of V. A persistence module V is called indecomposable if whenever V is expressed by a direct sum decomposition

$$V \simeq W \oplus W',$$

then $W = 0$ or $W' = 0$, i.e., V does not have nontrivial direct sum decompositions. Concerning decompositions of modules, the following theorem is important.

Theorem 1 (Krull-Remak-Schmidt) *A persistence module V can be expressed by a direct sum*

$$V \simeq W^{(1)} \oplus \cdots \oplus W^{(\ell)}$$

of indecomposable modules $W^{(1)}, \ldots, W^{(\ell)}$. In addition, if

$$V \simeq W^{(1)} \oplus \cdots \oplus W^{(\ell)} \simeq X^{(1)} \oplus \cdots \oplus X^{(m)},$$

where $W^{(1)}, \ldots, W^{(\ell)}$ and $X^{(1)}, \ldots, X^{(m)}$ are indecomposable, then $\ell = m$ and there exists a permutation σ on $\{1, \ldots, \ell\}$ such that $W^{(i)} \simeq X^{(\sigma(i))}$ for $i = 1, \ldots, \ell$.

Given a filtration of topological spaces $\mathbb{X} : X_1 \subset \cdots \subset X_n$, the persistent homology $H_\ell(\mathbb{X})$ of $\mathbb{X}$ is defined by a representation [3]

$$H_\ell(X_1) \longrightarrow H_\ell(X_2) \longrightarrow H_\ell(X_3) \longrightarrow \cdots\cdots \longrightarrow H_\ell(X_n)$$

on an A_n quiver

$$\overset{1}{\circ} \longrightarrow \overset{2}{\circ} \longrightarrow \overset{3}{\circ} \longrightarrow \cdots\cdots \longrightarrow \overset{n}{\circ}. \tag{5}$$

Here $H_\ell(X_i)$ is the homology group of X_i with field coefficient K, and so is a K-vector space located at the vertex i. The linear maps $H_\ell(X_i) \to H_\ell(X_{i+1})$ are induced by the inclusion maps $X_i \hookrightarrow X_{i+1}$.

Let $I(b, d)$ be a representation defined by a vector space

$$I(b,d)_i = \begin{cases} K, & b \le i \le d \\ 0, & \text{otherwise} \end{cases}$$

for each vertex i with identity maps between adjacent copies of K and zero maps otherwise. Then, it is easy to show that $I(b,d)$ is indecomposable. Gabriel's theorem [8] shows that, up to isomorphism, this gives a complete list of the indecomposable representations of (5).

Theorem 2 (Gabriel) *Any indecomposable representation V on a quiver* (5) *is isomorphic to $I(b,d)$ with some $1 \le b \le d \le n$.*

From Theorems 1 and 2, the persistent homology $H_\ell(\mathbb{X})$ can be uniquely expressed as

$$H_\ell(\mathbb{X}) \simeq \bigoplus_{1 \le b \le d \le n} I(b,d)^{n_{bd}},$$

where $n_{bd} \in \mathbb{N}_0$. The success of persistent homology relies on this unique decomposition theorem. In particular, persistent homology can be classified by interval representations $I(b,d)$, and is compactly expressed by the so-called persistence diagrams.

Definition 1 The persistence diagram $D_\ell(\mathbb{X})$ of $\mathbb{X}$ is a multiset of points in $\{(i,j) \mid i,j = 1,\dots,n,\ i \le j\}$ such that the multiplicity of (b,d) is given by n_{bd}.

From the unique decomposition property, persistent topological features in a filtration can be studied by the persistence diagrams.

We now generalize this scenario of persistent homology to representations on the commutative triple ladder. Let U and V be persistence modules on (3). A morphism $s : U \to V$ is called a section if there exists a left inverse morphism $s' : V \to U$ (i.e., $s' \circ s = 1_U$). Similarly, a morphism $r : U \to V$ is a retraction if there exists a right inverse morphism $r' : V \to U$ (i.e., $r \circ r' = 1_V$).

Definition 2 A morphism $f : U \to V$ is said to be irreducible if

1. f is neither a section nor a retraction,
2. if $f = g \circ h$, then either g is a retraction or h is a section.

This can be interpreted as the fact that irreducible morphisms are those which admit no nontrivial factorizations.

Let U and V be indecomposable persistence modules on (3). We denote by $\mathrm{rad}(U,V)$ the K-vector space of all non-invertible morphisms from U to V. Let us also denote by $\mathrm{rad}^2(U,V)$ the set of all morphisms of the form $f \circ g$, where $f \in \mathrm{rad}(W,V)$ and $g \in \mathrm{rad}(U,W)$ with a module W (not necessarily indecomposable). Then, $\mathrm{rad}^2(U,V)$ is a subspace of $\mathrm{rad}(U,V)$. Moreover, it can be shown that a morphism $f : U \to V$ is irreducible if and only if $f \in \mathrm{rad}(U,V) \setminus \mathrm{rad}^2(U,V)$ (see Appendix in [1] for details). Hence, the quotient vector space $\mathrm{rad}(U,V)/\mathrm{rad}^2(U,V)$ measures the number of irreducible morphisms between two indecomposable modules.

Definition 3 The Auslander-Reiten quiver Γ of the commutative triple ladder is defined as follows:

1. The vertices of Γ are all the isomorphism classes of indecomposable modules of (3).
2. For two indecomposable isomorphism classes $[M]$ and $[N]$, if $\dim(\mathrm{rad}(M,N)/\mathrm{rad}^2(M,N)) = \ell$, then ℓ directed arrows are assigned from $[M]$ to $[N]$.

Auslander-Reiten Quiver of Commutative Triple Ladder

In this section, we first provide the explicit form of the Auslander-Reiten quiver of the commutative ladder (3). Then, from the finiteness of the Auslander-Reiten quiver, a generalization of persistence diagrams for the commutative ladder (3) is naturally derived.

On the commutative triple ladder, let us denote a persistence module

$$\begin{array}{ccccc} K^{n_4} & \longrightarrow & K^{n_5} & \longleftarrow & K^{n_6} \\ \uparrow & & \uparrow & & \uparrow \\ K^{n_1} & \longrightarrow & K^{n_2} & \longleftarrow & K^{n_3} \end{array}$$

by $\begin{smallmatrix} n_4 n_5 n_6 \\ n_1 n_2 n_3 \end{smallmatrix}$. This is called its dimension vector expression. In the paper [7], the following theorem is presented.

Theorem 3 *The Auslander-Reiten quiver Γ of the commutative triple ladder* (3) *is given in Fig. 1. The morphisms on $K \to K$ are all identity maps, and the maps in* $\begin{smallmatrix} 111 \\ 121 \end{smallmatrix}$ *and* $\begin{smallmatrix} 121 \\ 010 \end{smallmatrix}$ *are given by*

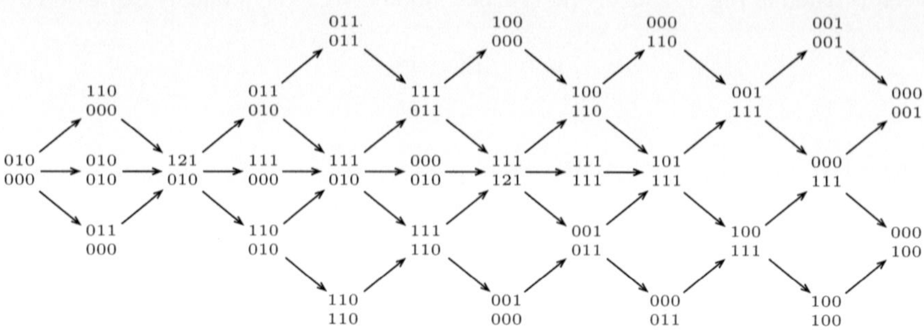

Fig. 1 The Auslander-Reiten quiver Γ of (3)

$$\begin{array}{ccccc} K & \longrightarrow & K & \longleftarrow & K \\ \uparrow & & {\scriptstyle f_{25}}\uparrow & & \uparrow \\ K & \xrightarrow[f_{12}]{} & K^2 & \xleftarrow[f_{32}]{} & K \end{array}, \quad f_{12} = \begin{pmatrix} 1 \\ 0 \end{pmatrix}, f_{32} = \begin{pmatrix} 0 \\ 1 \end{pmatrix}, f_{25} = \begin{pmatrix} 1 & 1 \end{pmatrix},$$

$$\begin{array}{ccccc} K & \xrightarrow{f_{45}} & K^2 & \xleftarrow{f_{65}} & K \\ \uparrow & & {\scriptstyle f_{25}}\uparrow & & \uparrow \\ 0 & \longrightarrow & K & \longleftarrow & 0 \end{array}, \quad f_{45} = \begin{pmatrix} 1 \\ 0 \end{pmatrix}, f_{65} = \begin{pmatrix} 0 \\ 1 \end{pmatrix}, f_{25} = \begin{pmatrix} 1 \\ 1 \end{pmatrix}.$$

Sketch of a proof For any indecomposable nonprojective module L and any indecomposable noninjective module M, there exist the so-called almost split sequences

$$0 \to \tau L \to X \to L \to 0, \tag{6}$$

$$0 \to M \to Y \to \tau^{-1} M \to 0, \tag{7}$$

where the functors τ and τ^{-1} are called the Auslander-Reiten translations (see IV.2 in [1]). In addition, τL and $\tau^{-1}M$ are indecomposable modules, and morphisms in these exact sequences consist of direct sums of irreducible morphisms into indecomposable modules.

Starting from nonprojective and noninjective indecomposable modules, the exact sequences (6) and (7) generate new indecomposable modules τL and $\tau^{-1}M$ with irreducible morphisms. By successively applying this procedure, we obtain a finite connected component of Γ shown in Fig. 1. Then, it is known that if Γ possesses a finite connected component Γ_0, then $\Gamma = \Gamma_0$ (see IV.4 in [1]). This completes the proof. □

From Theorem 3, all the indecomposable modules are classified by the dimension vectors listed in Fig. 1, and any persistence module V can be uniquely expressed by

$$V \simeq \bigoplus_{[M] \in \Gamma} [M]^{n_{[M]}}, \tag{8}$$

where $[M] \in \Gamma$ is a vertex in Γ and $n_{[M]} \in \mathbb{N}_0$. Hence, these dimension vectors are the complete discrete invariants for persistence modules on (3). In particular, the dimension vector $\begin{smallmatrix} 111 \\ 111 \end{smallmatrix}$ gives us information about robust common topological features.

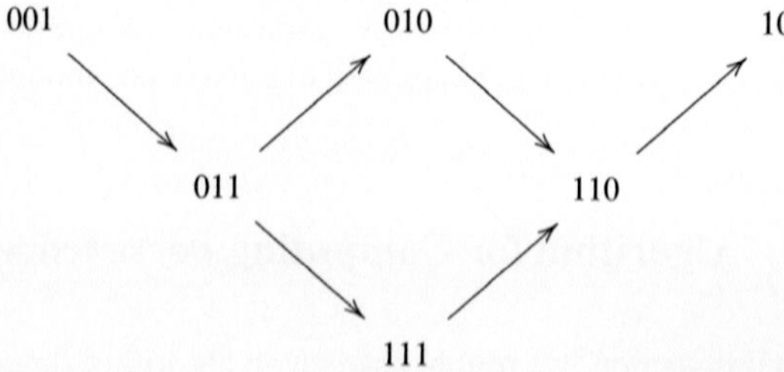

Fig. 2 The Auslander-Reiten quiver of (9)

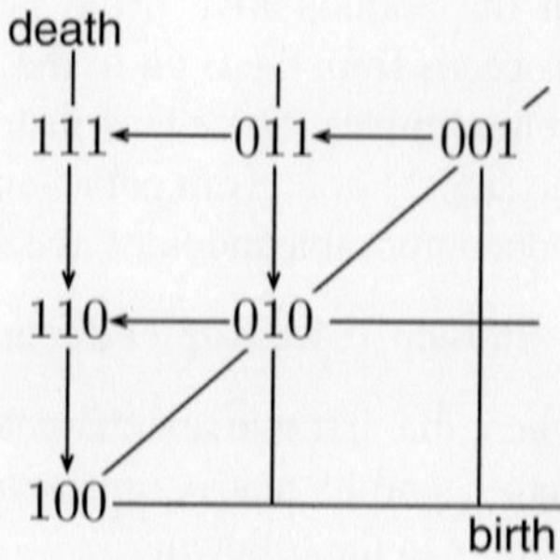

Fig. 3 Grid poins in persistence diagrams

In order to explain the relationship between the Auslander-Reiten quiver and persistence diagrams, let us consider the A_3 quiver

$$\overset{1}{\circ} \longrightarrow \overset{2}{\circ} \longrightarrow \overset{3}{\circ} \tag{9}$$

as an example. Its Auslander-Reiten quiver is given in Fig. 2 and can be obtained by using the same strategy as in the proof of Theorem 3.

By definition, each indecomposable module corresponds to a grid point in the persistence diagram shown in Fig. 3. This means that persistence diagrams can be seen as functions on the vertex set of the Auslander-Reiten quiver taking values in $\mathbb{N}_0$. Similarly, we have a correspondence between the Auslander-Reiten quiver and persistence diagrams on a general A_n quiver.

From this remark, the following is a natural generalization of persistence diagrams.

Definition 4 The persistence diagram $D(V)$ of a persistence module (8) on the commutative triple ladder is a function on the vertex set of the Auslander-Reiten quiver Γ:

$$D(V) = \{n_{[M]} \in \mathbb{N}_0 \mid [M] : \text{vertex of } \Gamma\}.$$

That is, the value of $D(V)$ at $[M]$ is given by $n_{[M]}$ in the indecomposable decomposition of V in (8).

Remark 1 This definition of persistence diagrams by means of the Auslander-Reiten quiver Γ can be applied to persistence modules on other quivers.

3 Algorithm for Computing Persistence Diagram

In this section, we roughly sketch an algorithm for computing persistence diagrams on the commutative triple ladder. The algorithm, which consists of five steps, proceeds from the right to the left in Fig. 4, which is obtained from the Auslander-Reiten quiver. The edges instead show how the indecomposables are grouped by the steps. For a given persistence module V, we successively derive corresponding indecomposable modules at each step.

In Step 1, we extract direct summands consisting of $\genfrac{}{}{0pt}{}{100}{100}$, $\genfrac{}{}{0pt}{}{000}{100}$, $\genfrac{}{}{0pt}{}{001}{001}$, and $\genfrac{}{}{0pt}{}{000}{001}$, where the first two are characterized by Kerf_{12}, and the rest by Kerf_{32}. Note that a map f_{ij} and its matrix representation using a certain basis are denoted by the same symbol in the following.

Algorithm (Step 1)

1. Perform a base change on V^1 to obtain a column echelon form

$$f_{12} = \left(0 \middle| A\right) : V^1 = K^1 \oplus L^1 \to V^2,$$

where the submatrix A is in a column echelon form with nonzero columns. The subspaces K^1 and L^1 are defined by this decomposition of the map f_{12}.

2. Perform a base change on V^4 to obtain a row echelon form

$$f_{14} = \left(\begin{array}{c|c} B & C \\ \hline 0 & D \\ 0 & 0 \end{array}\right) : V^1 = K^1 \oplus L^1 \to V^4 = V^4_{14} \oplus U^4,$$

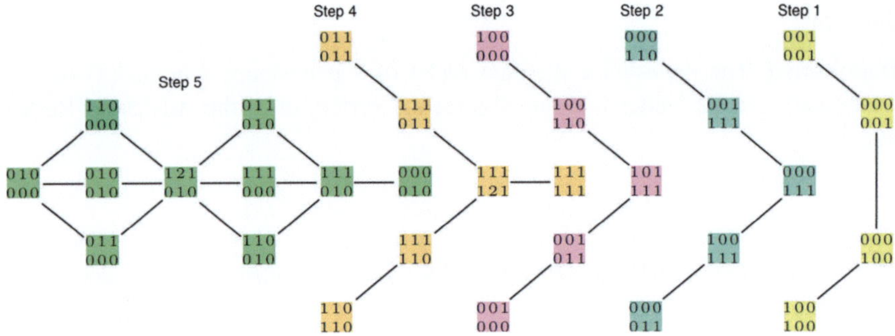

Fig. 4 Algorithm flow chart

where the submatrices B and D are in row echelon forms, and consist of nonzero rows. The subspace V^4_{14} is defined by the first component of the row decomposition, and U^4 is defined by the rest.

3. Perform base changes on K^1 and L^1, respectively, to transform f_{14} into

$$f_{14} = \left(\begin{array}{c|c|cc} 0 & 1 & E & 0 \\ \hline 0 & 0 & 0 & 1 \\ \hline 0 & 0 & 0 & 0 \end{array}\right) : V^1 = V^1_1 \oplus V^1_{14} \oplus L^1 \to V^4 = V^4_{14} \oplus U^4,$$

where V^1_1 and V^1_{14} are defined in a similar manner as above, and 1 is the identity matrix of a suitable size. This base change has an effect on f_{12} as

$$f_{12} = \left(\begin{array}{c|c|cc} 0 & 0 & \tilde{A}_1 & \tilde{A}_2 \end{array}\right) : V^1 = V^1_1 \oplus V^1_{14} \oplus L^1 \to V^2.$$

4. Perform a base change on $V^1 = V^1_1 \oplus V^1_{14} \oplus L^1$ by the base change matrix

$$\left(\begin{array}{c|c|cc} 1 & 0 & 0 & 0 \\ \hline 0 & 1 & -E & 0 \\ \hline 0 & 0 & 1 & 0 \\ 0 & 0 & 0 & 1 \end{array}\right)$$

to transform f_{14} into

$$f_{14} = \left(\begin{array}{c|c|c|c} 0 & 1 & 0 & 0 \\ \hline 0 & 0 & 0 & 1 \\ \hline 0 & 0 & 0 & 0 \end{array}\right) : V^1 = V^1_1 \oplus V^1_{14} \oplus U^1 \to V^4 = V^4_{14} \oplus U^4.$$

Note that this base change on V^1 does not change f_{12} at all.

From these procedures, we obtain the direct summands corresponding to $\begin{smallmatrix}000\\100\end{smallmatrix}$ and $\begin{smallmatrix}100\\100\end{smallmatrix}$ as

$$\begin{array}{ccccc} 0 & \longrightarrow & 0 & \longleftarrow & 0 \\ \uparrow & & \uparrow & & \uparrow \\ V^1_1 & \longrightarrow & 0 & \longleftarrow & 0 \end{array}, \qquad \begin{array}{ccccc} V^4_{14} & \longrightarrow & 0 & \longleftarrow & 0 \\ \uparrow & & \uparrow & & \uparrow \\ V^1_{14} & \longrightarrow & 0 & \longleftarrow & 0 \end{array}.$$

By symmetry, we can similarly obtain the direct summands corresponding to $\begin{smallmatrix}001\\001\end{smallmatrix}$ and $\begin{smallmatrix}000\\001\end{smallmatrix}$. □

The remaining steps are basically executed in a similar manner to Step 1. However, in Step 4 and Step 5, the indecomposable modules with two dimensional vector spaces require more detailed analysis. It should be noted that, in each step, we first execute the *subspace tracking* (1, 2 in Step 1), and then do the *basis arrangement* (3, 4 in Step 1). The role of the subspace tracking mode is to characterize the corresponding subspaces, while that of the basis arrangement mode is to separate these subspaces as direct summands by properly choosing bases. Also, let us remark that the process of subspace tracking can be regarded as deriving streamlined modules, as used in [3].

4 Numerical Example

In this section, we show a simple numerical example to detect robust common topological features between two geometric objects. The purpose is to illustrate the use of computation and interpretation of the persistence diagrams. Via a simple example, we build up intuition about how to read the persistence diagrams. For more detailed analysis (with complete algorithm) and practical applications, we refer to [7].

In our test example, we set eight points $P = \{p_i \in \mathbb{R}^3 \mid i = 1, \dots, 8\}$ on the vertices of a cube with length 4, and construct its α-complex X_r with a ball radius parameter r by CGAL [5]. We also set another eight points $Q = \{q_i \in \mathbb{R}^3 \mid i = 1, \dots, 8\}$ obtained by adding small perturbations on $p_i \in P$, and construct its α-complex Y_r. From the bijective correspondence $p_i \leftrightarrow q_i$ of the points in P and Q, we set horizontal maps

$$X_r \rightarrow X_r \cup Y_r \leftarrow Y_r$$

as simplicial maps into the simplicial complex $X_r \cup Y_r$. A different parameter value $s(> r)$ induces inclusions $X_r \hookrightarrow X_s$ and $Y_r \hookrightarrow Y_s$, and leads to a triple ladder geometric model as in (1). We study the persistence module obtained by taking its first homology $H_1(\bullet)$. Note that there exist 5 nonzero generators for $H_1(X_r)$ in the radius parameter $r \in [2, 2\sqrt{2})$.

Figures 5 and 6 are the persistence diagrams for $r = 2.052$, $s = 2.5$ and $r = 2.1$, $s = 2.5$, respectively. In the first set of parameters, we observe that there are two robust common topological features measured by the multiplicity at $\begin{smallmatrix}111\\111\end{smallmatrix}$. Furthermore, three remaining persistent homology generators in $H_1(X_r) \rightarrow H_1(X_s)$

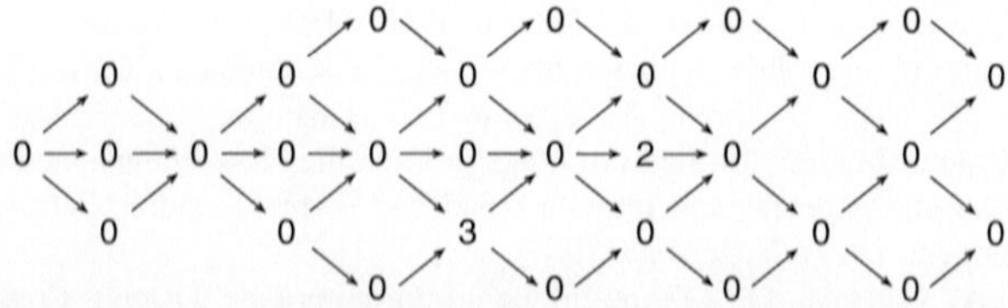

Fig. 5 Persistence diagram for $r = 2.052$, $s = 2.5$

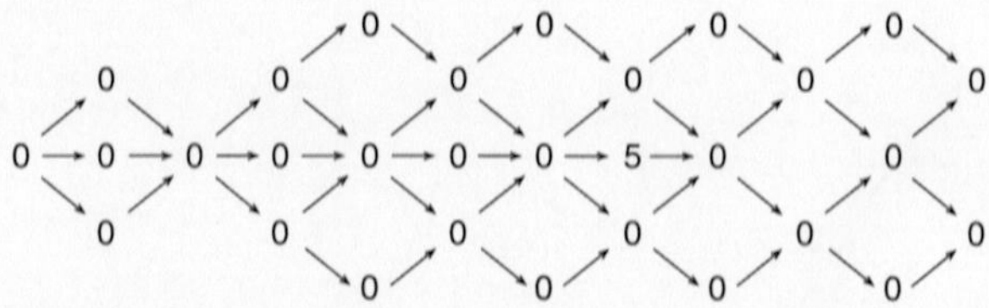

Fig. 6 Persistence diagram for $r = 2.1$, $s = 2.5$

are counted at $\begin{matrix}111\\110\end{matrix}$, and we infer from the position of the 0 that these generators are not shared by Y_r.

On the other hand, in the second set of parameters, all the persistent homology generators in $H_1(X_r) \to H_1(X_s)$ are shared with $H_1(Y_r) \to H_1(Y_s)$, since the multiplicity of $\begin{matrix}111\\111\end{matrix}$ is five.

Acknowledgements This work is partially supported by JSPS Grant-in-Aid and JST CREST.

References

1. Assem, I., Simson, D., Skowroński, A.: Elements of the Representation Theory of Associative Algebras 1: Techniques of Representation Theory. Cambridge University Press, Cambridge/New York (2006)
2. Auslander, M., Reiten, I., Smalø, S.: Representation Theory of Artin Algebras. Cambridge University Press, Cambridge (1995)
3. Carlsson, G., de Silva, V.: Zigzag persistence. Found. Comput. Math. **10**, 367–405 (2010)
4. Carlsson, G., Zomorodian, A.: The theory of multidimensional persistence. Discret. Comput. Geom. **42**, 71–93 (2009)
5. CGAL: Computational Geometry Algorithms Library, http://www.cgal.org/
6. Edelsbrunner, H., Letscher, D., Zomorodian, A.: Topological persistence and simplification. Discret. Comput. Geom. **28**, 511–533 (2002)
7. Escolar, E.G., Hiraoka, Y.: Persistence modules on commutative ladders of finite type. Discret. Comput. Geom. **55**, 100–157 (2016)
8. Gabriel, P.: Unzerlegbare Darstellungen I. Manuscr. Math. **6**, 71–103 (1972)
9. Gameiro, M., Hiraoka, Y., Izumi, S., Kramar, M., Mischaikow, K., Nanda, V.: A topological measurement of protein compressibility. Jpn. J. Indust. Appl. Math. **32**, 1–17 (2015)

10. Hiraoka, Y., Nakamura, T., Hirata, A., Escolar, E.G., Matsue, K., Nishiura, Y.: Hierarchical structures of amorphous solids characterized by persistent homology. arXiv:1501.03611
11. Nakamura, T., Hiraoka, Y., Hirata, A., Escolar, E.G., Nishiura, Y.: Persistent homology and many-body atomic structure for medium-range order in the glass. Nanotechnology **26**, 304001
12. Ringel, C.M.: Tame Algebras and Integral Quadratic Forms. Lecture Notes in Mathematics, vol. 1099. Springer, Berlin/New York (1984)
13. Zomorodian, A., Carlsson, G.: Computing persistent homology. Discret. Comput. Geom. **33**, 249–274 (2005)

Construction of Negatively Curved Cubic Carbon Crystals via Standard Realizations

Hisashi Naito

Dedicate to Yumiko Naito

Abstract In Tagami et al. (Carbon 76:266–274, 2014), we constructed *physically stable* sp2 negatively curved cubic carbon structures which reticulate a Schwarz P-like surface. The method for constructing such crystal structures is based on the notion of the standard realization of abstract crystal lattices. In this paper, we expound on the mathematical method to construct such crystal structures.

Keywords Discrete surfaces • Carbon structures • Schwarzites

1 Introduction

In the last few decades, there have been many studies about carbon allotropes. In this paper, we mainly study their geometric structures. For example, it is considered that C60, graphene sheets, and single wall carbon nanotubes (SWNTs) span 2-dimensional surfaces in $\mathbb{R}^3$. In particular, they span the sphere S^2, the plane $\mathbb{R}^2$ and cylinders, respectively. Since the Gauss curvature of S^2 is positive and that of $\mathbb{R}^2$ and cylinder is zero, C60 spans a positively curved surface and both of graphene sheets and SWNTs span flat surfaces.

It is natural to ask whether there is an sp2 carbon allotrope which spans a negatively curved surface. Mackay-Terrones [2] first constructed such an sp2 carbon crystal structure. In this paper, we call this "Mackay-Terrones C192". The structure can be placed on the Schwarz P-surface, which is a well-known triply periodic minimal surface. Since minimal surfaces have negative Gauss curvature, we may consider Mackay-Terrones C192 to be a negatively curved "discrete surface". Following Mackay-Terrones' work, there have been many examples of carbon crystal structures which can be considered to be triply periodic sp2 negatively curved discrete surfaces (cf. [3–9]). In [1], we also constructed such examples, and

H. Naito (✉)
Graduate School of Mathematics, Nagoya University, Nagoya, Japan
e-mail: naito@math.nagoya-u.ac.jp

Y. Nishiura, M. Kotani (eds.), *Mathematical Challenges in a New Phase of Materials Science*, Springer Proceedings in Mathematics & Statistics 166,
DOI 10.1007/978-4-431-56104-0_5

the main features of our method were mathematical and systematic. To construct such structures, we used the notion of the standard realization of topological crystals (cf. [10, 11]).

In this paper, first, in Sect. 2, we summarize the combinatorial structures of sp2 carbon allotropes and define the notion of "discrete surfaces" and their "total discrete curvature" . In Sect. 3, we expound the notion of topological crystals and their standard realizations. After that, in Sect. 4, we discuss how to construct such structures by using the standard realization of topological crystals.

The main purpose of this paper is to construct *physically stable* negatively curved sp2 carbon structures. In particular, we only consider structures that can be placed on a surface whose symmetry is the same as the Schwarz P-surface. We note that when we discuss the mathematical structure of sp2 "carbon" networks, atomic species do not effect the construction of such structures except for searching for physically stable structures by using first principle calculations.

2 Minimal Surfaces and Schwarzites

Minimal surfaces in $\mathbb{R}^3$ are examples of surfaces with negative curvature. Moreover, there are many examples of triply periodic minimal surfaces. One well-known example is the Schwarz P-surface [12–15] (cf. Fig. 1a), which has following properties:

1. It is a triply periodic minimal surface, whose fundamental domain is cubic,
2. It is parameterized by a conformal map from a Riemann surface with genus 3,
3. It has the same symmetry the **pcu** net. In other words, its space group is $Pm\bar{3}m$.

We remark that Schwarz D- and G-surfaces also satisfy properties 1 and 2, and the following discussion may apply to D- and G-surfaces. However, for simplicity, we only discuss the Schwarz P-surface by assuming property 3. Here, *cubic* means

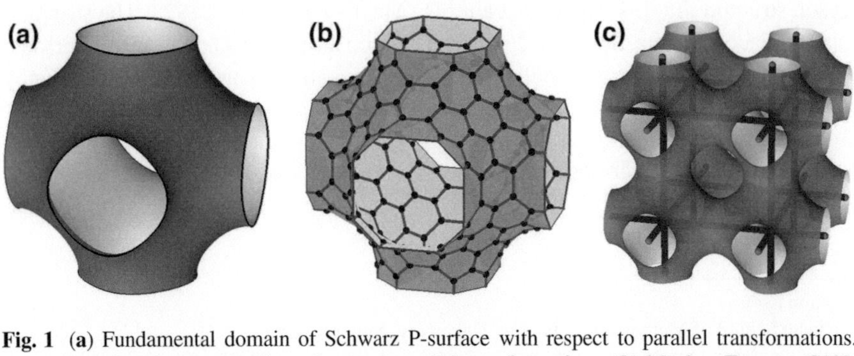

Fig. 1 (**a**) Fundamental domain of Schwarz P-surface with respect to parallel transformations. The total surface is expanded by orthogonal parallel transformations. (**b**) Mackay-Terrones C192, embedded on Schwarz P-surface. (**c**) Schwarz P-surface and **pcu** net

that the period lattice is orthogonal, that is to say, the gram matrix of the period lattice is proportional to the identity matrix. In [2], Mackay-Terrones construct a carbon crystal structure which is placed on the Schwarz P-surface (cf. Fig. 1b). Since the structure is placed on a triply periodic minimal surface and it is a crystal structure, hence, it can be considered to be an example of a "negatively curved carbon crystal". Here, we note that structures which are placed on a triply periodic minimal surface are called *schwarzites*, whereas structures which are placed on a positively curved surface are called *fullerenes*. More precisely, a schwarzite is a trivalent network (an sp2 structure) that reticulates a triply periodic hyperbolic surface. As we mention in Remark 2.3, such a structure contains rings larger than hexagons. Later, in [5], Lenosky et al. also constructed similar structures. There are many works constructing schwarzites (cf. [3, 4, 6, 7, 16–20]).

On the other hand, locally finite graphs are suitable mathematical objects to consider as molecular/crystal structures. We consider not only merely abstract graph structures but also realizations of graphs. As mentioned in the Introduction, since we will study surfaces which are spanned by molecular/crystal structures, we consider polyhedra constructed by realizations of graphs, and call them "discrete surfaces".

Definition 2.1 ([11, Section 7.1]) Let $X = (V, E)$ be a locally finite graph. A map $\Phi: X \longrightarrow \mathbb{R}^d$ is called a d-dimensional *realization* of X, identifying X with a 1-dimensional cell complex.

From this section, a realization X of a locally finite graph denotes a molecular/crystal structure, and we only consider 3-dimensional ones. Moreover, if the structure is periodic with period lattice Γ, we write X/Γ as X if there is no confusion. Hence, we only consider only realizations of a finite graphs to study molecular/crystal structures.

Now we define the notion of "discrete surfaces". Let X be a realization of a finite graph. Moreover, we assume each simple closed path of X spans a surfaces which does not self-intersect. Since such a realization is considered to be a polyhedron, we call X a *discrete surface*. In particular, if the underlying finite graph of X is of degree 3, we call X a *discrete surface of degree* 3.

Now, we consider C60, SWNTs and sp2 schwarzites. Expressing them by realizations of graphs, each underlying graph of X is of degree 3 (*trivalent*), since we consider only sp2 structures.

Let $V(X)$, $E(X)$ and $F(X)$ be the numbers of vertices, edges and faces in a discrete surface X. By considering X as a CW complex, we may define the genus $g(X)$ of X as $g(X) = \dim H_1(X, \mathbb{Z})$. Then, by Euler's theorem (Euler-Poincaré theorem),

$$V(X) - E(X) + F(X) = 2 - 2g(X). \tag{1}$$

For a smooth surface M, by the Gauss-Bonnet Theorem, $2\pi(2-2g) = K(M)$, where $K(M)$ is the total curvature of M. Therefore, we define the *total discrete curvature* $K(X)$ of X by

$$K(X) = V(X) - E(X) + F(X).$$

Table 1 C60 structure is a truncated icosahedron, For graphene sheets and SWNTs, we take fundamental domain with respect to $\mathbb{Z}^2$- and $\mathbb{Z}$-action, respectively. For Mackay-Terrones C192, we take the fundamental domain with respect to $\mathbb{Z}^3$-action. See Appendix for chiral index of SWNTs

	$V(X)$	$E(X)$	$F(X)$	$K(X)$	$g(X)$
C60	60	90	32	2	0
SWNT, chiral index = (6, 0)	24	36	12	0	1
SWNT, chiral index = (6, 3)	64	126	42	0	1
SWNT, chiral index = (6, 6)	24	36	12	0	1
Mackay-Terrones C192	192	288	102	−4	3

Here, we only consider the signature of $K(X)$, and we say that X is totally negatively curved if and only if $K(X) < 0$. Hence, we can call a carbon structure negatively curved if the realization X of the graph of the structure is negatively curved.

Example 2.2 C60 is totally positively curved, and the total discrete curvatures of SWNTs and graphene sheets are zero. Mackay-Terrones C192 is totally negatively curved (cf. Table 1).

Remark 2.3 For a discrete surface X of degree 3, we obtain

$$V(X) = \frac{k}{3}\sum_k N_k, \quad E(X) = \frac{k}{2}\sum_k N_k, \quad F(X) = \sum_k N_k, \tag{2}$$

where N_k is the number of k-gon. By using Euler theorem, we, therefore, obtain

$$\sum_k \left(1 - \frac{k}{6}\right) N_k = 2 - 2g(X). \tag{3}$$

If X satisfies $K(X) < 0$, Eq. (3) implies that X contains at least one k-gon ($k \geq 7$).

3 Topological Crystals and Their Standard Realizations

Scientists use space groups (crystallographic groups) to describe structures of crystals. A space group represents the symmetry of atoms in the crystal, but it does not describe bonds of atoms. The notion of a crystal lattice or a topological crystal, which was introduced by Kotani-Sunada [10], represents placements of atoms and their bonds in a crystal. The definition of topological crystals is as follows:

Definition 3.1 ([11, Section 6.2]) A locally finite graph $X = (V, E)$ is called a *topological crystal* or a *crystal lattice* if and only if there exists a finite graph

$X_0 = (V_0, E_0)$ and a regular covering map $\pi : X \longrightarrow X_0$ such that the covering transformation group Γ of π is abelian. Moreover, if the rank of $\Gamma \subset H_1(X_0, \mathbb{Z})$ is d, X is called d-dimensional.

A topological crystal $X = (V, E)$ is an abstract structure of a crystal, that is to say, each vertex of X represents an atom of a crystal, and if $\boldsymbol{v}_1$ and $\boldsymbol{v}_2 \in V$ are connected by an edge $(\boldsymbol{v}_1, \boldsymbol{v}_2) \in E$, then atoms $\boldsymbol{v}_1$ and $\boldsymbol{v}_2$ are bonded. However, the coordinates of vertices are not defined. To determine, the coordinates of vertices, we define a realization of X.

Definition 3.2 ([11, Section 7.1]) Let $X = (V, E)$ be a d-dimensional topological crystal, which is identified with a 1-dimensional cell complex. A map $\Phi : X \longrightarrow \mathbb{R}^d$ is called a *realization* of X. Moreover, if there exists an injective homomorphism $\rho : \Gamma \longrightarrow \mathbb{R}^d$ such that $\Phi(\sigma \boldsymbol{v}) = \Phi(\boldsymbol{v}) + \rho(\sigma)$ for any $\boldsymbol{v} \in V$ and $\sigma \in L$, and $\rho(\Gamma)$ is a lattice subgroup of $\mathbb{R}^d$, then Φ is called a *periodic* realization.

Since physical crystals have periodic structures, it is natural to consider only periodic realizations. Figure 2 shown examples of realizations of the same crystal lattices. As show in Fig. 2, we would like to select good ones among all realizations of a given crystal lattice. To do this, we define the energy functional with respect to realizations of a given crystal lattice.

Definition 3.3 ([11, Section 7.3]) The *energy* of a realization Φ of a crystal lattice X is defined by

$$E(\Phi) = \frac{1}{2} \sum_{(\boldsymbol{v}_1, \boldsymbol{v}_2) \in E_0} |\Phi(\boldsymbol{v}_1) - \Phi(\boldsymbol{v}_2)|^2,$$

where X_0 is the fundamental graph of X. Moreover, Φ is called a *harmonic* realization if Φ is a critical point of E.

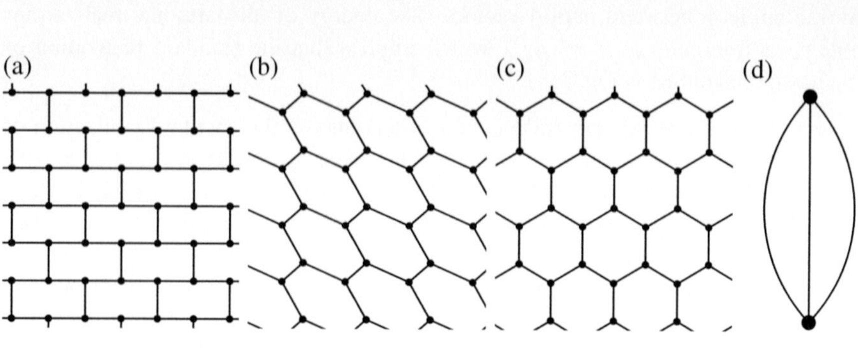

Fig. 2 Structures (**a**), (**b**) and (**c**) are periodic realizations of the same topological crystals (hexagonal lattice). (**a**) is not a harmonic realization, (**b**) and (**c**) are harmonic. Moreover (**c**) is the standard realization of the hexagonal lattice. (**d**) is the fundamental graph of the hexagonal lattice

It is easy to show that Φ is harmonic if and only if Φ satisfies

$$\sum_{(v,v_1)\in E_0} (\Phi(v) - \Phi(v_1)) = 0, \tag{4}$$

for all $v \in X_0$. Note that (1) the left-hand side of (4) is the discrete Laplacian of $(\Phi(V_0), E_0)$; (2) Φ is harmonic if and only if all vertices of $\Phi(X)$ satisfy mechanical equilibrium; (3) structures in Fig. 2b, c are harmonic. However, we consider the structure in Fig. 2c as the most symmetric object among all realizations of X. To select the most symmetric object by using the variational method, we also consider variations with respect to lattices Γ.

Definition 3.4 ([11, Section 7.4]) Let X be a d-dimensional crystal lattice with the fundamental graph X_0. Moreover, let Γ be the covering transformation group. A realization $\Phi\colon X \longrightarrow \mathbb{R}^d$ is called *standard* if and only if Φ is a critical point of E with respect to variations Φ and Γ subject to $\mathrm{Vol}(\Gamma) = 1$.

Theorem 3.5 ([10]) *For any topological crystal X, there exists a standard realization Φ of X. Moreover Φ is unique up to scaling and parallel translations.*

Remark 3.6 In [21, Theorem 4], Delgado-Friedrichs considered equilibrium placements $\Phi\colon X \to \mathbb{R}^d$ of d-periodic graphs X, and proved the unique (up to affine translations) existence of the equilibrium placement for any periodic graph (See also [22]). The notion of equilibrium placements corresponds to the notion of harmonic realizations. Moreover, Delgado-Friedrichs also proved that, for any d-periodic graph X, there exists a unique equilibrium placement Φ such that, for every $\gamma \in \Gamma \subset \mathrm{Aut}(X)$, an isometry $\gamma^*\colon \mathbb{R}^d \to \mathbb{R}^d$ associated with γ with respect to Φ exists ([21, Theorem 11]). This equilibrium placement corresponds to the standard realization.

Example 3.7 For the hexagonal lattice (Fig. 2), all lattices Γ are parameterized by the angle θ between period vectors. The energy of all harmonic realizations attains its minimum at $\theta = \pi/3$, which implies that the standard realization of the hexagonal lattice is Fig. 2c.

Example 3.8 There is a general method of calculating the standard realization of a given topological crystal (cf. [11, 23]). The diamond crystal is obtained by this procedure from the fundamental graph in Fig. 3a. The **srs** network, which has a deep relationship with the Gyroid surface, is also obtained by this method. Moreover, Sunada [24] shows that the only networks which satisfy the "strongly isotropic property" are diamond (**dia**) and **srs**. For the relation of the **srs** network and the Gyroid surface, see [25, 26]. For the definition of the strongly isotropic property, see [24, p.212], and for the explicit figure of K_4, see [23], [26, Figure 5]. Here, we note that we show physical meta-stability of carbon K_4 structure in [27].

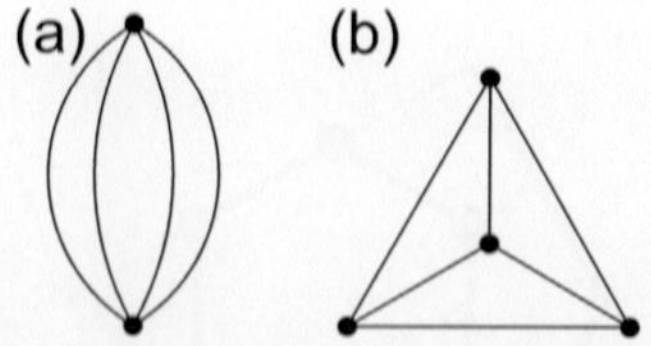

Fig. 3 (**a**) The fundamental graph of the diamond lattice. The diamond lattice is obtained by the standard realization of the maximal abelian covering of this graph. (**b**) The fundamental graph (K_4 graph) of the **srs** net (the K_4 lattice). The **srs** net is obtained by the standard realization of the maximal abelian covering of K_4 graph

The most important property of the standard realization is as follows:

Theorem 3.6 ([24, Theorem 1]) *Let X be a d-dimensional crystal lattice, and Φ a realization of X. Then there exists a homomorphism $T\colon \mathrm{Aut}(X) \longrightarrow M(d)$ such that $\Phi(gx) = T(g)\Phi(x)$ for any $x \in X$ and $g \in \mathrm{Aut}(X)$, where $\mathrm{Aut}(X)$ is the automorphism group of X and $M(d)$ is the motion group of $\mathbb{R}^d$.*

Theorem 3.6 implies that the standard realization is the most symmetric among all realizations.

4 Construction of Negatively Curved Carbon Crystals

Our main aim is to construct examples of physically stable sp2 negatively curved carbon structures. In particular, we construct such structures with the same symmetry as **pcu**. For the 3-net **pcu**, see Delgado-Friedrichs et al. [28] and Hyde et al. [26]. In the following, we abbreviate the symmetry as *cubic symmetry*. In this section, we use the word *network*, which means a realization of graphs.

Construction of Topological Crystals

To construct network (graph) structures with cubic symmetry, first we construct them in the *hexagonal region*, then extend to the fundamental region (unit cell) of crystals (see Figs. 1a and 4a). Since the hexagonal region has reflective symmetry of order 3, we only consider networks with such symmetry. We call the fundamental region of such symmetry the *kite-region* (see Fig. 4b). Using the orbifold notation, the kite-region is the $*2223$ orbifold, and the hexagonal region is the $*222222$ orbifold in $\mathbb{H}^2$. For the orbifold notation, see [29, 30]

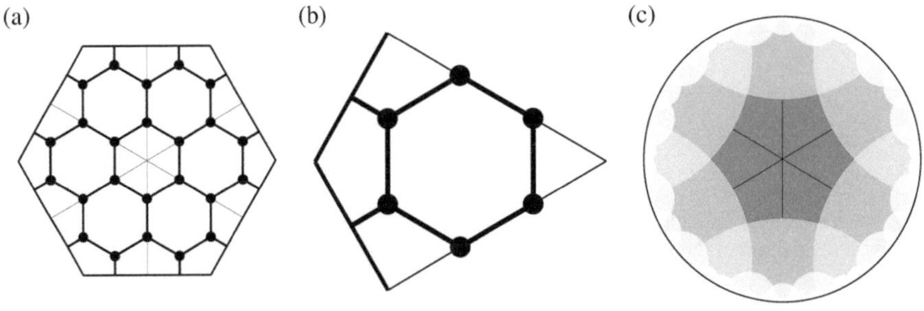

Fig. 4 (**a**) The combinatorial structure of Mackay-Terrones C192 in the hexagonal region (∗222222). Mackay-Terrones C192 is obtained by reflecting this structure by group action of ∗222222. (**b**) The combinatorial structure of Mackay-Terrones C192 in the kite-region (∗2223). Structure (**a**) is obtained by reflecting the structure in the kite-region with respect to lines through the right vertex (the vertex with $\pi/3$ symmetry). (**c**) An image of the hexagonal region in $\mathbb{H}^2$ (See [31, Figure 7(c)])

As the first step to construct such networks, we construct networks in the kite-region satisfying the following properties:

1. Any inner vertex is of degree 3,
2. Any vertex on the boundary is joined with the two neighbouring vertices on the boundary, or with an inner vertex and not with both neighbouring vertices on the boundary,
3. A network is planar and connected, and there are at least four vertices on the boundary,
4. A network does not have a consecutive sequence of odd vertices on the boundary,
5. A network is triangle-free.

Condition 1 corresponds to consideration of sp2 crystals. By conditions 2, 3 and 4, we may extend the network to the hexagonal region. Condition 5 implies avoiding 3-rings in structures. It is easy to prove that networks with the above conditions must have an even number of vertices. Networks with 6 and 8 vertices in the kite-region are classified in [1, Figure 2]. Since the kite-region is homeomorphic to the disk, we may construct such networks in the disk region.

The second step is to extend networks in the kite-region to the hexagonal region by reflections. In this step, if there exist degree 2 vertices on reflection boundaries, we delete them. By these procedures, we may construct a network in the hexagonal region.

Finally, by extending the network in the hexagonal region to the fundamental region by parallel transformations, we obtain a required network. We note that the above properties are not sufficient conditions to obtain trivalent networks which reticulate a Schwarz P-like surface. Constructing networks in the hexagonal region, and patching them in a suitable manner, we may obtain networks on D- and/or G-like surface. However, as the aim of this note is to construct examples of networks on a P-like surface, we consider the hexagonal region as a hexagonal face of the trucated octahedron, then we obtain a required trivalent network by extending in the above manner.

Construction of Standard Realizations

Let $X_0 = (V_0, E_0)$ be a network constructed as in Sect. 4. Since X_0 has cubic symmetry, we may extend X_0 to a 3-dimensional toplogical crystal $X = (V, E)$. Our purpose in this section is to construct the standard realization of X, and to prove that the realization has cubic symmetry. That is to say, translation vectors of the standard realization are orthogonal.

First we let $\Phi: X \longrightarrow \mathbb{R}^3$ be a realization with $\Gamma = X/X_0$, and $\{\boldsymbol{e}_x, \boldsymbol{e}_y, \boldsymbol{e}_z\}$ be the basis of $\mathbb{R}^3$ with $\det(\boldsymbol{e}_x\boldsymbol{e}_y\boldsymbol{e}_z) = 1$. By using this realization, we may find coordinates of vertices $\boldsymbol{x}_i = \Phi(\boldsymbol{v}_i) \in \mathbb{R}^3$ for $\boldsymbol{v}_i \in V$. To calculate standard realizations, we should define the energy of realizations. If $\boldsymbol{v}_i \in V_0$, then edges $(\boldsymbol{v}_i, \boldsymbol{v}_j)$ satisfy either $\boldsymbol{v}_j \in V_0$, $\boldsymbol{v}_j \in V_0 + \boldsymbol{e}_\alpha$ or $\boldsymbol{v}_j \in V_0 - \boldsymbol{e}_\alpha$ for some $\alpha \in \{x, y, z\}$, hence, we write that $(\boldsymbol{v}_i, \boldsymbol{v}_j)$ is in either E_0, $E_{(0,+)}$ or $E_{(0,-)}$. Under this notation, the energy of Φ may be written as

$$E = \frac{1}{2}\left(\sum_{(\boldsymbol{v}_i,\boldsymbol{v}_j)\in E_0} |\boldsymbol{x}_i - \boldsymbol{x}_j|^2 \right. \\ \left. + \sum_{(\boldsymbol{v}_i,\boldsymbol{v}_j)\in E_{(0,+)}} |\boldsymbol{x}_i + \boldsymbol{e}_\alpha - \boldsymbol{x}_j|^2 + \sum_{(\boldsymbol{v}_i,\boldsymbol{v}_j)\in E_{(0,-)}} |\boldsymbol{x}_i - \boldsymbol{e}_\alpha - \boldsymbol{x}_j|^2 \right). \tag{5}$$

Since the standard realization is the critical point of E with variations with respect to variables $\{\boldsymbol{x}_i\}$ and $\{\boldsymbol{e}_\alpha\}$ subject to $\det(\boldsymbol{e}_x\boldsymbol{e}_y\boldsymbol{e}_z) = 1$, then to obtain the standard realization, using Lagrangian multiplier, we must solve the following equations:

$$\frac{\partial E}{\partial \boldsymbol{x}_i} = 0, \tag{6}$$

and

$$\frac{\partial}{\partial \boldsymbol{e}_\alpha}(E - \lambda \det(\boldsymbol{e})) = 0. \tag{7}$$

Proposition 4.1 ([1, Theorem 2 of Supplementary materials]) *The linear system* (6) *is solvable.*

Proof The left hand side of Eq. (6) is

$$\frac{\partial E}{\partial \boldsymbol{x}_i} = \boldsymbol{x}_{j_1} + \boldsymbol{x}_{j_2} + \boldsymbol{x}_{j_3} - 3\boldsymbol{x}_j + \boldsymbol{b}_j \tag{8}$$

where $\{\boldsymbol{v}_{j_1}, \boldsymbol{v}_{j_2}, \boldsymbol{v}_{j_3}\}$ are vertices adjacent to $\boldsymbol{v}_j$, and $\boldsymbol{b} = (\boldsymbol{b}_i)$ is defined by

$$\boldsymbol{b}_i = \begin{cases} +\boldsymbol{e}_\alpha & \text{if } (\boldsymbol{v}_j, \boldsymbol{v}_{j\ell}) \in E_{(\alpha,+)} \text{ for some } \ell \in \{1,2,3\}, \alpha \in \{x,y,z\}, \\ -\boldsymbol{e}_\alpha & \text{if } (\boldsymbol{v}_j, \boldsymbol{v}_{j\ell}) \in E_{(\alpha,-)} \text{ for some } \ell \in \{1,2,3\}, \alpha \in \{x,y,z\}, \\ 0 & \text{otherwise} \end{cases}$$

That is to say, the Eq. (6) is written as

$$\Delta_G \boldsymbol{x} = \boldsymbol{b}, \tag{9}$$

where $\Delta_G = \boldsymbol{A} - 3\boldsymbol{I}$ is the discrete Laplacian of $G = X/\Gamma$, and A is the adjacency matrix of G. Since the discrete Laplacian of a connected graph has only a 1-dimensional kernel, and the kernel is spanned by $(1, \ldots, 1)$. Since $\sum \boldsymbol{b}_i = 0$, we obtain that $\boldsymbol{b}$ is perpendicular to Ker Δ_G. hence, Eq. (6) is solvable. □

Theorem 4.2 ([1, Theorem 3 of Supplementary materials]) *The gram matrix of the lattice* $\{\boldsymbol{e}_x, \boldsymbol{e}_y, \boldsymbol{e}_z\}$ *which gives the solution of* (6) *and* (7) *is the identity matrix.*

Proof Let $\{\boldsymbol{x}_i\}$ be the solution of (6) with $\sum \boldsymbol{x}_i = 0$, and t_α be the reflection with respect to the plane with normal vector $\boldsymbol{e}_\alpha$. Moreover, let T be the group generated by $\{t_x, t_y, t_z\}$. Since the energy E is invariant under the action of T, the solution x satisfies:

1. If $T_x(\boldsymbol{v}_i) = \boldsymbol{v}_j$, then $\boldsymbol{x}_i = K_{i,x}\boldsymbol{e}_x + K_{i,y}\boldsymbol{e}_y + K_{i,z}\boldsymbol{e}_z$ and $\boldsymbol{x}_j = -K_{i,x}\boldsymbol{e}_x + K_{i,y}\boldsymbol{e}_y + K_{i,z}\boldsymbol{e}_z$.
2. $(\boldsymbol{v}_i, \boldsymbol{v}_j) \in E_{(\alpha,+)}$ if and only if $(\boldsymbol{v}_j, \boldsymbol{v}_i) \in E_{(\alpha,-)}$.

Therefore we obtain

$$\frac{\partial E}{\partial \boldsymbol{e}_\alpha} = \sum_{(\boldsymbol{v}_i,\boldsymbol{v}_j)\in E_{(\alpha,+)}} (\boldsymbol{x}_j + \boldsymbol{e}_\alpha - \boldsymbol{x}_i) - \sum_{(\boldsymbol{v}_i,\boldsymbol{v}_j)\in E_{(\alpha,-)}} (\boldsymbol{x}_j - \boldsymbol{e}_\alpha - \boldsymbol{x}_i) = K_\alpha \boldsymbol{e}_\alpha.$$

Since the action exchanging $\boldsymbol{e}_x$ and $\boldsymbol{e}_y$ belongs to T, we may obtain $K := K_x = K_y = K_z$ and $K \neq 0$. By 2, we obtain $K\boldsymbol{e}_\alpha = \lambda \boldsymbol{e}_\beta \times \boldsymbol{e}_\gamma$ with $\epsilon_{\alpha\beta\gamma} = \epsilon_{xyz}$. Hence, we obtain that $K\langle \boldsymbol{e}_\alpha, \boldsymbol{e}_\beta \rangle = \lambda \delta_{\alpha\beta}$. □

Remark 4.3 Theorem 4.2 is also obtained from Theorem 3.6. By Theorem 3.6, the standard realization has maximal symmetry, and the symmetry of the realized crystal must be same as that of the topological crystal. Since the topological crystal, which we are considering, is invariant under triply periodicity and cubic group action, therefore, such actions extend to actions $M(d)$, Hence, the gram matrix of $\{\boldsymbol{e}_\alpha\}$ should be proportional to the identity matrix.

Construction of Stable Configurations

By using Theorem 4.2, we obtain candidates for negatively curved cubic crystals. However, distances of neighbouring atoms in such structures are not almost the same. By physico-chemical considerations, the distances should be almost the same for physically stable configurations. To find stable configurations with respect to binding energies, we perturb coordinates of atoms, then find stable ones, which we call *relaxed configuration*, by using the first principle calculations. For this purpose, setting the standard realization as the initial configuration, perturbing positions of atoms, and calculating binding energies of carbon structures, we obtain stable configurations. By using the above method, we search for physically stable configuration, and we obtain negatively curved cubic sp2 carbon crystals as examples of schwarzites which reticulate a Schwarz P-like surface (Fig. 5).

Result 4.4 *We obtain four physically stable structures:* 6-1-1-P (C176), 6-1-2-P (C152), 6-1-3-P (C152) *and* 8-4-2-P (C168). *Relaxed configurations of them are physically stable. Moreover,* 6-1-1-P (C176), 6-1-2-P (C152), 6-1-3-P (C152) *are metals and* 8-4-2-P (C168) *is a semi-conductor.*

Our structures are illustrated in Fig. 6, and Table 2 lists their basic properties. For their energy bands and phonon spectrum of them, see [1, Figures 4–8].

Remark 4.5 To calculate relaxed configuration, the only information we use is the coordinates of carbon atoms. In Fig. 6d, we draw edges between an atom and the nearest three atoms. As a result, each relaxed configuration has the same topological structure as the original standard realization.

Hence, our method to obtain cubic schwarzites is effective.

Remark 4.6 In [31], Ramsden-Robins-Hyde discuss 3-dimensional Euclidean nets (3-net). Our examples are also 3-nets, and there are many examples which reticulate such surfaces. However, we construct, in particular, physically stable sp2 (trivalent) examples.

Remark 4.7 We note that the structure *6-1-2-P* (C152) is almost the same as with C152 in Park et al. [32].

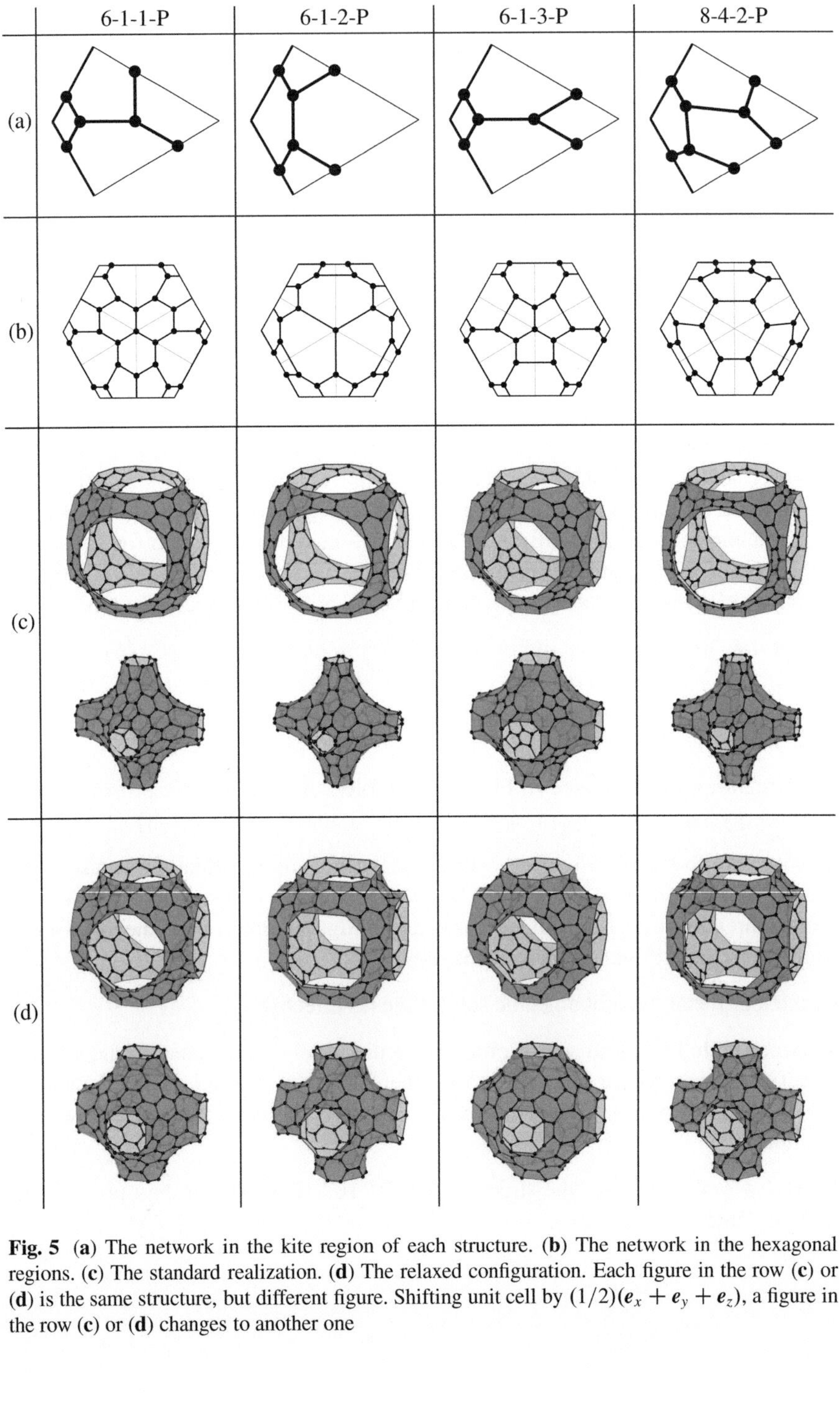

Fig. 5 (**a**) The network in the kite region of each structure. (**b**) The network in the hexagonal regions. (**c**) The standard realization. (**d**) The relaxed configuration. Each figure in the row (**c**) or (**d**) is the same structure, but different figure. Shifting unit cell by $(1/2)(\boldsymbol{e}_x + \boldsymbol{e}_y + \boldsymbol{e}_z)$, a figure in the row (**c**) or (**d**) changes to another one

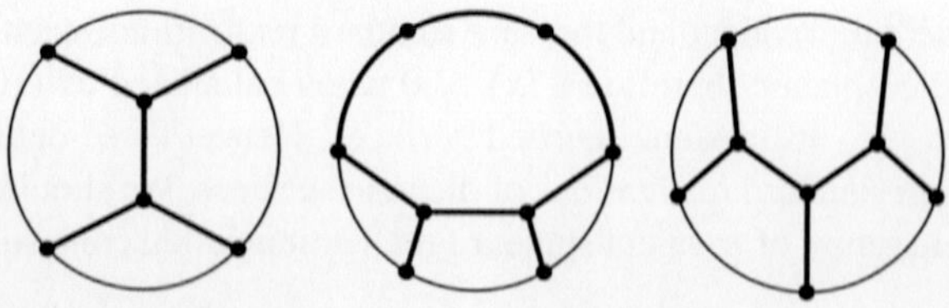

Fig. 6 Networks in circle region. See [1, Figure 2] for other networks with 6- and 8-vertices. Mackay-Terrones C192 can be constructed from 8-2. (**a**) 6-1. (**b**) 8-2. (**c**) 8-4

Table 2 Number of k-gon N_k of our structures. Each structure lies on a negatively curved surface, and has $g = 3$. Hence, all of these structures are negatively curved. Obviously, each structure satisfies (3) for $g = 3$

Schwarzites	N_5	N_6	N_7	N_8	$V(X)$	$E(X)$	$F(X)$	$K(X)$	$g(X)$
6-1-1-P (C176)	0	60	24	0	176	264	84	−4	3
6-1-2-P (C152)	0	60	24	0	152	240	84	−4	3
6-1-3-P (C152)	24	12	24	12	152	228	72	−4	3
8-4-2-P (C168)	0	68	0	12	168	252	80	−4	3
Mackay-Terrones C192	0	90	0	12	192	288	102	−4	3

5 Further Problems

Finally, we note further problems about negatively curved carbon structures.

In addition to the Schwarz P-surface, there are well-known triply periodic minimal surfaces, namely, the Schwarz D-surface and G-surface (gyroid surface), and there are many studies about sp2 carbon schwarzites related to D- and G-surfaces, for example [3, 5, 25, 33, 34]. In this paper, we discuss trivalent networks, which reticulate the P-surface, and construct examples of physically stable carbon sp2 structures. Now, we are interested in the following problem: Whether we can construct examples of physically stable carbon sp2 structures which reticulate D- or G-surfaces using our method? For example, calculate the physical stability of the carbon sp2 structures that is constructed from 6-1-1, 6-1-2, 6-1-3 or 8-4-2 and reticulates the D- or G-surface.

In this note, we claim that a discrete surface is negatively curved, if and only if the total discrete curvature is negative. In other words, a discrete surface is negatively curved if the structure is placed on a surface with negative curvature. The first definition is not a point-wise characterization of negativity, and the last one is not rigorous. On the other hand, there are definitions of curvature for general discrete surfaces. For examples, [35–37]. One well-known definition of curvature is

$$K(\boldsymbol{x}) = 2\pi - \sum_{j=0}^{k-1} \theta_j, \quad \theta_j = \angle \boldsymbol{x}_j \boldsymbol{x} \boldsymbol{x}_{j+1} \tag{10}$$

where $\{x_j\}_{j=0}^{k-1}$ are the neighbours of x (see [14, 37]). However, since the structures that we consider are trivalent and they are standard realizations, each vertex and its neighbours are co-planer, therefore $K(x) \equiv 0$ when calculated by (10). Definitions of curvature in [35, 36] assume meshed surfaces. Hence, these definitions do not apply to trivalent standard realizations of discrete surfaces. We should consider how to define the curvature of trivalent standard realizations of discrete surfaces.

Acknowledgements The author gratefully thanks to Prof. Davide M. Proserpio who informed us about articles on this subject and that 6-1-2-P is almost the same structure as C152 in [32]. The author also thanks Professor Motoko Kotani and Professor Yasumasa Nishiura. They gave the author an opportunity to talk at the Symposium "Mathematical Challenge to a New Phase of Materials Science" in Kyoto, 2014. The author was partially supported by Grants-in-Aid for Scientific Research (C) 40211411 and (A) 15H02055.

Dedication I dedicate this paper to my late wife Yumiko Naito (September 8 1963–October 2 2015). While battling with breast cancer, she spent her days with a positive and enthusiastic attitude, supporting my research and caring for our son. My research up to now could not have existed without her support. I dedicate this paper to Yumiko's support during 24 years of marriage and to the wonderful memories of the 30 years since we first met. To my wife Yumiko, with gratitude and love.

Appendix: Indexes of Single Wall Nanotubes

We summarize the geometric structure of single wall nanotubes (SWNTs). Mathematically, a SWNT is considered as the fundamental region of the $\mathbb{Z}$-action on the standard realization of the hexagonal lattice. In the followings, we explain the geometric structures of SWNTs.

First, we define

$$\begin{aligned} \boldsymbol{v}_0 &= (0,0), \quad \boldsymbol{v}_1 = (-\sqrt{3}/2, 1/2), \quad \boldsymbol{v}_2 = (\sqrt{3}, 1/2), \quad \boldsymbol{v}_3 = (0,-1), \\ \boldsymbol{a}_1 &= \boldsymbol{v}_2 - \boldsymbol{v}_1 = (\sqrt{3}, 0), \quad \boldsymbol{a}_2 = \boldsymbol{v}_3 - \boldsymbol{v}_1 = (\sqrt{3}/2, -3/2), \end{aligned}$$

then the graph $X_0 = (V_0, E_0)$, $V_0 = \{\boldsymbol{v}_i\}_{i=0}^3$, $E_0 = \{(\boldsymbol{v}_0, \boldsymbol{v}_i)\}_{i=1}^3$ is the fundamental region of the hexagonal lattice, and $\{\boldsymbol{a}_1, \boldsymbol{a}_2\}$ is the basis of the parallel transformations (see Fig. 7). Here, the angle between $\boldsymbol{a}_1$ and $\boldsymbol{a}_2$ is $\pi/3$. By using the basis, we may write all vertices $\boldsymbol{v}$ of the hexagonal lattice as

$$\boldsymbol{v} = \alpha_1 \boldsymbol{a}_1 + \alpha_2 \boldsymbol{a}_2, \quad (\alpha_1, \alpha_2) \in \mathbb{Z} \times \mathbb{Z}, \quad \text{or} \quad (\alpha_1, \alpha_2) \in (\mathbb{Z} - 1/3) \times (\mathbb{Z} - 1/3).$$

Definition The vector $\boldsymbol{c} = c_1 \boldsymbol{a}_1 + c_2 \boldsymbol{a}_2$ is called the *chiral vector* and (c_1, c_2) is called the *chiral index* if and only if the SWNT is constructed from the hexagonal lattice by identifying $\boldsymbol{x}$ and $\boldsymbol{x} + \boldsymbol{c}$.

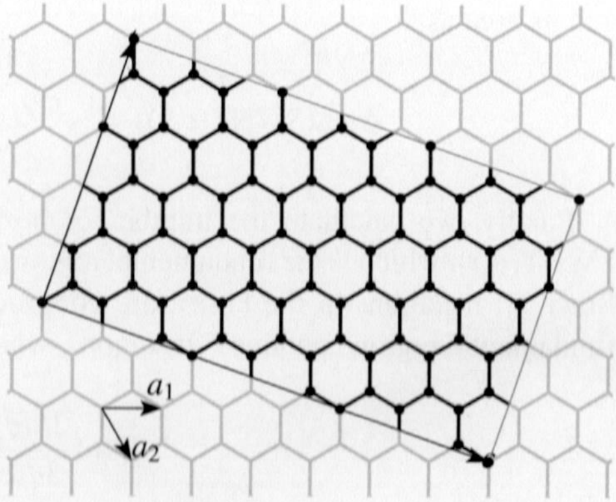

Fig. 7 Configuration of the single-wall carbon nanotube, with $\boldsymbol{c} = (6, 3)$ and $\boldsymbol{t} = (4, -5)$

The vector $\boldsymbol{t} = t_1\boldsymbol{a}_1 + t_2\boldsymbol{a}_2$, where $(t_1, t_2) = ((c_1 + 2c_2)/d(c), -(2c_1 + c_2)/d(\boldsymbol{c}))$, and $d(\boldsymbol{c}) = \gcd(c_1 + 2c_2, 2c_1 + c_2)$, is called the *lattice vector* of the SWNT with chiral index $\boldsymbol{c} = (c_1, c_2)$ (see Fig. 7).

The chiral vector indicates the direction of the circle of the SWNT, and is the period vector of the $\mathbb{Z}$-action on the hexagonal lattice. The chiral vector is orthogonal to the lattice vector, and the lattice vector is the minimum period of the hexagonal lattice along the tube axis of the SWNT. The circumference $L(\boldsymbol{c})$ of the SWNT is derived from the chiral index by $L(\boldsymbol{c}) = |\boldsymbol{c}| = \sqrt{3(c_1^2 + c_2^2 + c_1c_2)}$.The electronic properties of single wall carbon nanotubes depends on the chiral index (see [38]).

Next, let us consider finite length single wall nanotubes. More precisely, consider a finite length SWNT, which terminates at the vertices (atoms) with $\boldsymbol{x}_0 = \boldsymbol{0}$ and $\boldsymbol{x}_1 = \alpha_1\boldsymbol{a}_1 + \alpha_2\boldsymbol{a}_2$, and characterize the length of this SWNT by α_1 and α_2. In the followings, SWNT$(\boldsymbol{c}, \alpha_1\boldsymbol{a}_1 + \alpha_2\boldsymbol{a}_2)$ denotes the SWNT which is terminated by $\boldsymbol{0}$ and $\alpha_1\boldsymbol{a}_1 + \alpha_2\boldsymbol{a}_2$ and whose chiral vector is $\boldsymbol{c}$. It is easy to show that the length of SWNT$(\boldsymbol{c}, \boldsymbol{x})$ is $|\langle \boldsymbol{x}, \boldsymbol{e}_t\rangle|$, where $\boldsymbol{e}_t = \boldsymbol{t}/|\boldsymbol{t}|$. In [39], we define the *length index* $\ell(\boldsymbol{c}, \boldsymbol{x})$ of SWNT$(\boldsymbol{c}, \boldsymbol{x})$ as

$$\ell(\boldsymbol{c}, \boldsymbol{x}) = |\langle \boldsymbol{x}, \boldsymbol{e}_t\rangle| / \sqrt{3} = \frac{\sqrt{3}|c_2\alpha_1 - c_1\alpha_2|}{2\sqrt{c_1^2 + c_2^2 + c_1c_2}}, \quad \boldsymbol{x} = \alpha_1\boldsymbol{a}_1 + \alpha_2\boldsymbol{a}_2.$$

The length index corresponds to how many hexagons are arranged in the direction of the tube axis.

Now, we consider SWNT$(\boldsymbol{c}, \boldsymbol{t})$, which has the canonical length for the given chiral index $\boldsymbol{c}$. By using the definition of $\boldsymbol{t}$, we obtain the length index $\ell(\boldsymbol{c}, \boldsymbol{t})$ of SWNT$(\boldsymbol{c}, \boldsymbol{t})$ as

$$\ell(\boldsymbol{c}, \boldsymbol{x}) = \frac{\sqrt{3(c_1^2 + c_2^2 + c_1c_2)}}{d(\boldsymbol{c})} = \frac{L(\boldsymbol{c})}{d(\boldsymbol{c})},$$

and its area as

$$\text{Area}(\text{SWNT}(\boldsymbol{c},\boldsymbol{t})) = \sqrt{3}\ell(\boldsymbol{c},\boldsymbol{t})L(\boldsymbol{c}) = \frac{\sqrt{3}L(\boldsymbol{c})}{d(\boldsymbol{c})}.$$

Finally, we calculate the number of hexagons in the fundamental region of SWNT($\boldsymbol{c},\boldsymbol{t}$), which is the fundamental region of action generated by the lattice $\{\boldsymbol{c},\boldsymbol{t}\}$. Since all hexagons in the lattice are congruent and their volume are $3\sqrt{3}/2$, the fundamental region contains F hexagons, where

$$F = \frac{2L(\boldsymbol{c})^2}{3d(\boldsymbol{c})}. \tag{11}$$

Combining (2) and (11), we obtain the number of vertices V, of edges E, and of faces F in Table 1.

Note about figures Part of Figs. 6 and 1a are gray scale version of figures published in [1]. Figures 1a, b, 4 and 6c, d of 6-1-1-P are also published in [40].

References

1. Tagami, M., Liang, Y., Naito, H., Kawazoe, Y., Kotani, M.: Negatively curved cubic carbon crystals with octahedral symmetry. Carbon **76**, 266–274 (2014). http://dx.doi.org/10.1016/j.carbon.2014.04.077
2. Mackay, A.L., Terrones, H.: Diamond from graphite. Nature **352**, 762 (1991). http://dx.doi.org/10.1038/352762a0
3. Terrones, H., Mackay, A.L.: Triply periodic minimal-surfaces decorated with curved graphite. Chem. Phys. Lett. **207**, 45–50 (1993). http://dx.doi.org/10.1016/0009-2614(93)85009-d
4. Terrones, H., Mackay, A.L.: Negatively curved graphite and triply periodic minimal-surfaces. J. Math. Chem. **15**, 183–195 (1994). http://dx.doi.org/10.1007/bf01277558
5. Lenosky, T., Gonze, X., Teter, M., Elser, V.: Energetics of negatively curved graphitic carbon. Nature **355**, 333–335 (1992). http://dx.doi.org/10.1038/355333a0
6. Phillips, R., Drabold, D.A., Lenosky, T., Adams, G.B., Sankey, O.F.: Electronic-structure of schwarzite. Phys. Rev. B **46** 1941–1943 (1992). http://dx.doi.org/10.1103/PhysRevB.46.1941
7. Huang, M.Z., Ching, W.Y., Lenosky, T.: Electronic-properties of negative-curvature periodic graphitic carbon surfaces. Phys. Rev. B **47**, 1593–1606 (1993). http://dx.doi.org/10.1103/PhysRevB.47.1593
8. Townsend, S.J., Lenosky, T.J., Muller, D.A., Nichols, C.S., Elser, V.: Negatively curved graphitic sheet model of amorphous-carbon. Phys. Rev. Lett. **69**, 921–924 (1992). http://dx.doi.org/10.1103/PhysRevLett.69.921
9. Mackay, A.L., Terrones, H., Fowler, P.W.: Hypothetical graphite structures with negative gaussian curvature. Philos. Trans. R. Soc. A **343**, 113–127 (1993). http://dx.doi.org/10.1098/rsta.1993.0045
10. M. Kotani Sunada, T.: Standard realizations of crystal lattices via harmonic maps. Trans. Am. Math. Soc. **353**, 1–20 (2001).http://dx.doi.org/10.1090/S0002-9947-00-02632-5
11. Sunada, T.: Topological Crystallography. Surveys and Tutorials in the Applied Mathematical Sciences, vol. 6. Springer, Tokyo (2013). http://dx.doi.org/10.1007/978-4-431-54177-6

12. Schoen, A.H.: Infinite periodic minimal surfaces without self-intersections. NASA technical note, TN D-5541 (1970)
13. Schoen, A.H.: Reflections concerning triply-periodic minimal surfaces. Interface Focus **2**, 658–668 (2012). http://dx.doi.org/10.1098/rsfs.2012.0023
14. Karcher, H., Polthier, K.: Construction of triply periodic minimal surfaces. Philos. Trans. Roy. Soc. London Ser. A **354**, 2077–2104 (1996). http://dx.doi.org/10.1098/rsta.1996.0093
15. Molnar, E.: On triply periodic minimal balance surfaces. Struct. Chem. **13**, 267–275 (2002). http://dx.doi.org/10.1023/A:1015855721911
16. Rosato, V., Celino, M., Gaito, S., Benedek, G.: Thermodynamic behavior of a carbon schwarzite. Comput. Mater. Sci. **20**, 387–393 (2001). http://dx.doi.org/10.1016/S0927-0256(00)00197-X
17. Homyonfer, M., Feldman, Y., Margulis, L., Tenne, R.: Negative curvature in inorganic fullerene-like structure. Fullerene Sci. Tech. **6**, 59–66 (1998). http://dx.doi.org/10.1080/10641229809350185
18. Vanderbilt, D., Tersoff, J.: Negative-curvature fullerene analog of C60. Phys. Rev. Lett. **68**, 511–513 (1992). http://dx.doi.org/10.1103/PhysRevLett.68.511
19. Ceulemans, A., King, R.B., Bovin, S.A., Rogers, K.M., Troisi, A., Fowler, P.W.: The heptakisoctahedral group and its relevance to carbon allotropes with negative curvature. J. Math. Chem. **26**, 101–123 (1999). http://dx.doi.org/10.1023/A:1019129827020
20. King, R.B.: Chemical applications of topology and group theory .29. low density polymeric carbon allotropes based on negative curvature structures. J. Phys. Chem. **100**, 15096–15104 (1996). http://dx.doi.org/10.1021/jp9613201
21. Delgado-Friedrichs, O.: Equilibrium placement of periodic graphs and convexity of plane tilings. Discret. Comput. Geom. **33**, 67–81 (2005). http://dx.doi.org/10.1007/s00454-004-1147-x
22. Delgado-Friedrichs, O., O'Keeffe, M.: Identification of and symmetry computation for crystal nets. Acta Crystallogr. A **59**, 351–360 (2003). http://dx.doi.org/10.1107/S0108767303012017
23. Naito, H.: Visualization of standard realized crystal lattices. In: Spectral Analysis in Geometry and Number Theory. Contemporary Mathematics, vol. 484, pp. 153–164. American Mathematical Society, Providence (2009). http://dx.doi.org/10.1090/conm/484/09472
24. Sunada, T.: Crystals that nature might miss creating. Notices Am. Math. Soc. **55**, 208–215 (2008). Correction: "Crystals that nature might miss creating", ibid. **55**, 343 (2008)
25. Hyde, S.T., Ramsden, S.: Polycontinuous morphologies and interwoven helical networks. Europhys. Lett. **50**, 135–141 (2000). http://dx.doi.org/10.1209/epl/i2000-00245-y
26. Hyde, S.T., O'Keeffe, M., Proserpio, D.M.: A short history of an elusive yet ubiquitous structure in chemistry, materials, and mathematics. Angew. Chem. Int. Edit. **47**, 7996–8000 (2008). http://dx.doi.org/10.1002/anie.200801519
27. Itoh, M., Kotani, M., Naito, H., Sunada, T., Kawazoe, Y., Adschiri, T.: New metallic carbon crystal. Phys. Rev. Lett. **102**, 055703 (2009). http://dx.doi.org/10.1103/PhysRevLett.102.055703
28. Delgado-Friedrichs, O., O'Keeffe, M.O., Yaghi, O.M.: Three-periodic nets and tilings: semiregular nets. Acta Crystallogr. A **59**, 515–525 (2003). http://dx.doi.org/10.1107/S0108767303017100
29. Conway, J.H., Huson, D.H.: The orbifold notation for two-dimensional groups. Struct. Chem. **13**, 247–257 (2002). http://dx.doi.org/10.1023/A:1015851621002
30. Australian National University: EPINET Project. http://epinet.anu.edu.au/
31. Ramsden, S.J., Robins, V., Hyde, S.T.: Three-dimensional Euclidean nets from two-dimensional hyperbolic tilings: kaleidoscopic examples. Acta Crystallogr. A **65**, 81–108 (2009). http://dx.doi.org/10.1107/S0108767308040592
32. Park, S., Kittimanapun, K., Ahn, J.S., Kwon, Y.-K., Tománek, D.: Designing rigid carbon foams. J. Phys. Condens. Matter **22** (2010). http://dx.doi.org/10.1088/0953-8984/22/33/334220
33. Schroder, G.E., Ramsden, S.J., Christy, A.G., Hyde, S.T.: Medial surfaces of hyperbolic structures. Eur. Phys. J. B **35**, 551–564 (2003). http://dx.doi.org/10.1140/epjb/e2003-00308-y

34. Spagnolatti, I., Bernasconi, M., Benedek, G.: Electron-phonon interaction in carbon schwarzites. Eur. Phys. J. B **32**, 181–187 (2003). http://dx.doi.org/10.1140/epjb/e2003-00087-5
35. Müller, C., Wallner, J.: Oriented mixed area and discrete minimal surfaces. Discret. Comput. Geom. **43**, 303–320 (2010). http://dx.doi.org/10.1007/s00454-009-9198-7
36. Bobenko, A.I., Pottmann, H., Wallner, J.: A curvature theory for discrete surfaces based on mesh parallelity. Math. Ann. **348**, 1–24 (2010). http://dx.doi.org/10.1007/s00208-009-0467-9
37. Meyer, M., Desbrun, M., Schröder, P., Barr, A.H.: Discrete differential-geometry operators for triangulated 2-manifolds. In: Visualization and Mathematics III. Mathematics and Visualization, pp. 35–57. Springer, Berlin (2003)
38. Saito, R., Fujita, M., Dresselhaus, G., Dresselhaus, M.S.: Electronic structure of chiral graphene tubules. Appl. Phys. Lett. **60**, 2204–2206 (1992). http://dx.doi.org/10.1063/1.107080
39. Matsuno, T., Naito, H., Hitosugi, S., Sato, S., Kotani, M., Isobe, H.: Geometric measures of finite carbon nanotube molecules: a proposal for length index and filling indexes. Pure Appl. Chem. **86**, 489–495 (2014). http://dx.doi.org/10.1515/pac-2014-5006
40. Naito, H.: Chemistry and mathematics – discrete geometry and carbon structures. Mathe. Sci. (SUURI KAGAKU) **624**, 42–47 (2015). SAIENSU-SHA CO., LTD., (in Japanese)

A Note on a Local Ergodic Theorem for an Infinite Tower of Coverings

Ryokichi Tanaka

Abstract This is a note on a local ergodic theorem for a symmetric exclusion process defined on an infinite tower of coverings, which is associated with a finitely generated residually finite amenable group.

Keywords Hydrodynamic limit • Exclusion process • Tower of coverings

1 Introduction

The *hydrodynamic limit* is one of the main frameworks to understand scaling limit of interacting particle systems in order to capture the relation between microscopic and macroscopic phenomena in statistical physics. There have been studied a number of models such as exclusion processes (e.g., [4] and [9]), the stochastic Ginzburg-Landau model (e.g., [5]), a chain of anharmonic oscillators [13] and stochastic energy exchange models [14]. See also [8] and [16] for a background of this problem and its history. While a large number of studies have been devoted to understand stochastic models on the one-dimensional or the d-dimensional discrete torus and their scaling limits, several attempts have been made to generalise the underlying graphs to the ones with rich geometric structure. For example, Jara has studied a zero-range process on the Sierpinski gasket [6] and obtained a nonlinear heat equation which the limit of density empirical measure satisfies. The author gave a unified framework to discuss exclusion processes on a general class of graphs and obtained the hydrodynamic limit result for crystal lattices [17]. Recently, Sasada has given a unified and simplified approach to obtain the hydrodynamic limit for an important class of exclusion processes, so-called of *non-gradient* type, in general settings [15]. Although it is tempting to study a number of stochastic models on general underlying graphs from viewpoint of physics, the attempts made so far depend on the models and also on the structure of graphs. This note proposes a

R. Tanaka (✉)
Advanced Institute for Materials Research and Mathematical Institute, Tohoku University,
2-1-1 Katahira, Aoba-ku, Sendai, 980-8577, Japan
e-mail: rtanaka@wpi-aimr.tohoku.ac.jp

Y. Nishiura, M. Kotani (eds.), *Mathematical Challenges in a New Phase of Materials Science*, Springer Proceedings in Mathematics & Statistics 166,
DOI 10.1007/978-4-431-56104-0_6

possible approach to unify these generalisations on crystal lattices and self-similar graphs, and gives a *local ergodic theorem* for exclusion processes for a step toward to obtain the scaling limit.

In group theory, there is a remarkable class of groups which have been studied in the context of dynamical systems. Bartholdi, Grigorchuk and Nekrashevych described a number of self-similar (fractal) sets as the scaling limits of finite graphs associated with group actions [1]. They showed that a class of Julia sets and the Sierpinski gasket can be obtained as the scaling limits of *Schreier* graphs of some groups. See also [7] for another (yet related) construction of such a limit. These examples appear as groups acting on rooted trees, and a sequence of Schreier graphs is associated with the action on tree ([12] for a general background on this topic). Here we discuss the simplest case, an *infinite tower of coverings* associated with those kind of groups (Sect. 2). The group we consider is *finitely generated, infinite, residually finite amenable* (see Sect. 2 for the definition). We define an exclusion process on each finite covering graph, and take a limit as the size of graphs goes to infinity.

A limit theorem we are considering is a local ergodic theorem which describes local equilibrium states of particle systems. The theorem is also referred as the *replacement lemma*, e.g., in [8] and [17], and enables us to replace a local average of number of particles by a global average after a large enough time. There involves an ergodic theoretic argument to exchange the space average and the time average. In the hydrodynamic limit, this theorem has a role to describe several different local equilibrium states, and to verify the derivation of a (possibly non-linear) partial differential equation by pasting together those states. Here we formulate a local ergodic theorem by using a notion of *local function bundle* introduced in [17], and show it in the form of super exponential estimate (Theorem 2.1). The proof is based on the entropy method due to [5] and also [10]. In order to obtain the scaling limit result, one has to describe a limiting space, which we will not discuss in this note. The problem we discuss here is the first step before taking the scaling limit. It would be interesting to complete the second step and to obtain the hydrodynamic limit.

The organisation of this note is the following. In Sect. 2, we introduce the setting and examples of infinite tower of coverings, define exclusion processes and formulate a local ergodic theorem (Theorem 2.1). In Sect. 3, we show Theorem 2.1 and also prove auxiliary results: the one-block estimate (Theorem 3.2) and the two-blocks estimate (Theorem 3.3).

2 Notation and Results

Groups and Associated Infinite Towers of Coverings

Let $\Gamma = \langle S \rangle$ be a finitely generated group, where S is a finite symmetric set of generators, i.e., $S = S^{-1}$. Throughout this note Γ is an infinite group. We assume that Γ is *residually finite*, i.e., there is a descending sequence of finite index normal subgroups $\{\Gamma_i\}_{i=0}^{\infty}$ such that $\Gamma_i \lhd \Gamma$ with $[\Gamma : \Gamma_i] < \infty$ for each i, and $\bigcap_{i=1}^{\infty} \Gamma_i = \{\mathrm{id}\}$. We also assume that Γ is *amenable*, i.e., there exists a sequence of finite subsets $\{F_i\}_{i=0}^{\infty}$ of Γ such that $\lim_{i\to\infty} |\partial_S F_i|/|F_i| = 0$ where $\partial_S F_i := F_i S \setminus F_i$. We call such a sequence $\{F_i\}_{i=0}^{\infty}$ a *Følner sequence* and each F_i a *Følner set*. Henceforth we fix some finite generating set S in Γ and denote by $|x|_\Gamma$ the corresponding word norm of x in Γ, i.e., the minimum number of elements in S to obtain x.

Example 2.1

(i) *The group of integers* $\mathbb{Z}$. We can take $S = \{-1, 1\}$ as a finite symmetric set of generators. Considering the subgroups $\Gamma_i := 2^i\mathbb{Z}$ for positive integers i, one can check that the group is residually finite. The sets $F_i := [-i, i] \subset \mathbb{Z}$ gives a Følner sequence. In the same way, $\mathbb{Z}^d (d \geq 1)$ is also residually finite and amenable.
(ii) *Grigorchuck group.* Grigorchuck group is residually finite and a subsequence of balls forms a Følner sequence since it has subexponential volume growth. (See, e.g., Chapter 6 in [3].)
(iii) *Basilica group.* Basilica group is realized as a finitely generated subgroup in the automorphism group of the binary tree, and this implies that the group is residually finite. The group is also amenable as proved in [2].

Let $X = (V, E)$ be a Cayley graph of Γ associated with S, i.e., the set of vertices V is Γ, and the set of oriented edges $E := \{(x, xs) \in \Gamma \times \Gamma \ : \ s \in S\}$. Here we consider the Cayley graph as an oriented graph where both possibilities of orientation are included, i.e., if $e \in E$, then its reversed edge $\overline{e} \in E$. Denote the origin of e by oe and the terminus by te. The group Γ acts on X from the left hand side freely and vertex transitively. The quotient graph $X_0 := \Gamma\backslash X$ consists of one vertex and (oriented) loop edges. We use the graph distance d in X, and denote by $B(x, r)$ the ball in V centred at x with radius r. Abusing the notation, we denote the graph distance by d in other graphs as well.

Remark 2.1 We can extend our results to quasi-transitive graphs, i.e., Γ acts on X with a finite number of orbits. If $\Gamma = \mathbb{Z}^d$, the quasi-transitive graph X is called a *crystal lattice* ([11] and [17]).

Let $\{\Gamma_i\}_{i=0}^{\infty}$ be a descending sequence of normal subgroups such that $\Gamma_i \lhd \Gamma$ and $[\Gamma : \Gamma_i] < \infty$ for each i, and $\bigcap_{i=1}^{\infty} \Gamma_i = \{\mathrm{id}\}$. For each i, define the quotient finite graph $X_i := \Gamma_i\backslash X$. Each X_i is a finite graph since Γ_i is a finite index subgroup of Γ. Here Γ/Γ_i acts on X_i freely and vertex transitively, where Γ/Γ_i is a finite group. Then we have an infinite tower of coverings of finite graphs: $X_i \to X_0$ for $i = 1, 2, \ldots$.

Let us define a particle system on X. Define the configuration space by $Z := \{0, 1\}^V$, and denote each configuration by $\eta := \{\eta_x\}_{x\in V}$. The action of Γ lifts on Z naturally, by setting $(\sigma\eta)_z := \eta_{\sigma^{-1}z}$ for $\sigma \in \Gamma$, $\eta \in Z$ and $z \in V$. In the same way, for each quotient finite graph $X_m = (V_m, E_m) = \Gamma_m\backslash X$, we define a configuration space $Z_m := \{0, 1\}^{V_m}$. The action of Γ/Γ_m on X_m lifts on Z_m as above. Here we define a *local function bundle*, which is defined on the product space of the state space V and the configuration space Z. This is used to formulate our local ergodic theorem.

Definition 2.1 A Γ-invariant local function bundle for vertices is a function $f : V \times Z \to \mathbb{R}$ such that:

- There exists $r \geq 0$ such that for every $x \in V$, the function $f_x : Z \to \mathbb{R}$ depends only on $\{\eta_z\}_{z \in B(x,r)}$.
- For every $\sigma \in \Gamma, x \in V, \eta \in Z$, it holds that $f(\sigma x, \sigma\eta) = f(x, \eta)$.

In a similar way, a Γ-invariant local function bundle for edges is a function $f : E \times Z \to \mathbb{R}$ such that:

- There exists $r \geq 0$ such that for every $e \in E$, the function $f_e : Z \to \mathbb{R}$ depends only on $\{\eta_z\}_{z \in B(oe,r) \cup B(te,r)}$.
- For all $\sigma \in \Gamma, e \in E, \eta \in Z$, it holds that $f(\sigma e, \sigma\eta) = f(e, \eta)$.

Example 2.2

(i) For $x \in V$, if we define $f_x(\eta) := \eta_x$, then f is a Γ-invariant local function bundle for vertices. In this case, for each $x \in V$, the function f_x depends only on a configuration on x. The f is Γ-invariant by definition.
(ii) For $x \in V$, if we define $f_x(\eta) := \prod_{e \in E_x} \eta_{te}$, where $E_x := \{e \in E \ : \ oe = x\}$, then f is a Γ-invariant local function bundle for vertices.
(iii) For $e \in E$, if we define $f_e(\eta) := \eta_{oe} + \eta_{te}$, then f is a Γ-invariant local function bundle for edges.
(iv) For $e \in E$, if we define $f_e(\eta) := \prod_{e' \in E_{oe}} \eta_{te'} + \prod_{e' \in E_{te}} \eta_{te'} + c$, where $c > 0$, then f is a Γ-invariant local function bundle for edges and satisfies $f(e, \eta) \geq c$ for all $e \in E, \eta \in Z$.

Let $\{F_i\}_{i=1}^{\infty}$ be a Følner sequence for Γ. For a Γ-invariant local function bundle for vertices $f : V \times Z \to \mathbb{R}$, we define a *local average* associated with $\{F_i\}_{i=1}^{\infty}$. For $x \in V$ and F_i, let

$$\overline{f}_{x,i} := \frac{1}{|F_i|} \sum_{\sigma \in F_i} f_{\sigma x}.$$

Note that $\overline{f}_{\cdot,i}$ is again a Γ-invariant local function bundle.

Let us introduce a local average on a Følner set controlled by its size. For a non negative real number K, we define

$$b(K) := \max\{i \ : \ F_i \subseteq B(o, K)\},$$

i.e., $b(K)$ is the largest number L such that all $F_i, i = 1, \dots, L$ are included in the ball centred at o with radius K. Note that $b(K)$ is non-decreasing and goes to ∞ as

K goes to ∞. We also consider a local average of the following type:

$$\bar{f}_{x,b(K)} := \frac{1}{|F_{b(K)}|} \sum_{\sigma \in F_{b(K)}} f_{\sigma x}.$$

Fix a distinguished vertex $o \in V$. Now Γ acts on V transitively, so every vertex x can be written in the form $x = \gamma o$ for some $\gamma \in \Gamma$. For every $m \geq 1$, denote by the same character $o \in V_m$, the image of o via the covering map $X \to X_m$. For $\sigma \in \Gamma$, denote by $\underline{\sigma} \in \Gamma/\Gamma_m$ the image of σ via the canonical surjection. Since f is Γ-invariant, the Γ-invariant local function bundles f for vertices and edges induce functions on $V_m \times Z_m$, and $E_m \times Z_m$ respectively, they are Γ/Γ_m-invariant under the diagonal action for each m. We also use the same character for these induced ones.

Particle Systems

Assume that Z_m and Z are equipped with the prodiscrete topology, i.e., the product of discrete topology. Denote by $\mathcal{P}(Z_m)$ and by $\mathcal{P}(Z)$ the spaces of Borel probability measures on Z_m and on Z, respectively. We also define the $(1/2)$-Bernoulli measures ν_m on Z_m and ν on Z, as the direct product of the $(1/2)$-Bernoulli measures on $\{0, 1\}$.

Let us define the symmetric exclusion process on X. For a configuration $\eta \in Z$ and an edge $e \in E$, define by η^e a configuration exchanging states on oe and te, i.e.,

$$\eta^e_z := \begin{cases} \eta_{te} & z = oe, \\ \eta_{oe} & z = te, \\ \eta_z & \text{otherwise}, \end{cases}$$

for $z \in V$. Note that $\eta^e = \eta^{\bar{e}}$. Furthermore, for a function F on Z and $e \in E$, we define $\pi_e F(\eta) := F(\eta^e) - F(\eta)$. We also define the corresponding ones for $\eta \in Z_m$ and $e \in E_m$ in the same way. The *symmetric exclusion process* on X is defined in terms of a Γ-invariant local function bundle for edges. We call a Γ-invariant local function bundle $c : E \times Z \to \mathbb{R}$ is a *jump rate* if it satisfies the following properties:

- (symmetric) $c(e, \eta) = c(\bar{e}, \eta)$ and $c(e, \eta) = c(e, \eta^e)$ for all $e \in E, \eta \in Z$.
- (non-degenerate) There exists a positive constant c_0 such that $c_0 \leq c(e, \eta)$ for all $e \in E, \eta \in Z$.

Define the generator of a symmetric exclusion process by the operator $L_m : L^2(Z_m, \nu_m) \to L^2(Z_m, \nu_m)$,

$$L_m F(\eta) := \sum_{e \in E_m} c(e, \eta) \pi_e F(\eta),$$

for $F \in L^2(Z_m, \nu_m)$.

We consider a family of exclusion processes on an infinite tower of coverings. Each process on a covering graph X_m is speeded up by some time scaling factor t_m. Suppose that $\{t_m\}_{m=1}^{\infty}$ is an increasing sequence of positive numbers: $t_1 < t_2 < \cdots < t_m \to \infty$ with the condition $\sqrt{t_m} \le 2diamX_m$, where $diamX_m$ denotes the diameter of X_m.

Consider the continuous time Markov chain generated by L_m speeded up by t_m. Fix an arbitrary time $T > 0$ and denote by $D([0,T], Z_m)$ the space of paths being right continuous and having left limits. For a probability measure μ_m on Z_m, we define by $\mathbb{P}_m$ the distribution on $D([0,T], Z_m)$ of the continuous time Markov chain $\eta_m(t)$ generated by $t_m L_m$ with the initial measure μ_m.

For $0 \le \rho \le 1$, denote by ν_ρ the ρ-Bernoulli measure on Z, that is the direct product of the ρ-Bernoulli measures on $\{0,1\}$. We define a *global average* of a Γ-invariant local function bundle $f : V \times Z \to \mathbb{R}$ by $\langle f_o \rangle(\rho) := \mathbb{E}_{\nu_\rho}[f_o]$ the expectation of the function $f_o : Z \to \mathbb{R}$ with respect to ν_ρ.

The following estimate enables us to approximate a global average of a local function bundle by a local average of one under a suitable time-space average. This estimate is referred as a local ergodic theorem, which we show in the super exponential estimate.

Theorem 2.1 *Fix* $T > 0$. *For every* Γ*-invariant local function bundles* $f : V \times Z \to \mathbb{R}$ *and every* $\delta > 0$,

$$\lim_{i\to\infty} \limsup_{\varepsilon\to 0} \limsup_{m\to\infty} \frac{1}{[\Gamma : \Gamma_m]} \log \mathbb{P}_m \left(\frac{1}{[\Gamma : \Gamma_m]} \int_0^T V_{o,m,\varepsilon,i}(\eta(t))dt \ge \delta \right) = -\infty,$$

where

$$V_{o,m,\varepsilon,i}(\eta) = \sum_{\underline{\sigma}\in\Gamma/\Gamma_m} \left| \bar{f}_{\underline{\sigma}o,i}(\eta) - \langle f_o \rangle \left(\overline{\eta}_{\underline{\sigma}o,b(\varepsilon\sqrt{t_m})} \right) \right| .$$

Here we remark that this theorem generalises [17][Theorem 4.1] for crystal lattices.

3 Proof of Theorem 2.1

The super exponential estimate for $\mathbb{P}_m$ is reduced to $\mathbb{P}_m^{eq}$ which is the distribution of continuous time Markov chain generated by $t_m L_m$ with the initial measure the (1/2)-Bernoulli measure ν_m, i.e., an equilibrium measure. Indeed, for every measurable set $A \subset D([0,T], Z_m)$, $\mathbb{P}_m(A) \le 2^{[\Gamma:\Gamma_m]}\mathbb{P}_m^{eq}(A)$ and the factor $2^{[\Gamma:\Gamma_m]}$ does not contribute to the super exponential estimate. (Note that $[\Gamma : \Gamma_m] = |V_m|$.) Moreover, the super exponential estimate Theorem 2.1 is reduced to the following theorem.

For $\mu \in \mathcal{P}(Z_m)$, define the Dirichlet form for $\varphi := d\mu/d\nu_m$ by

$$I_m(\mu) := -\int_{Z_m} \sqrt{\varphi} L_m \sqrt{\varphi} d\nu_m,$$

which is also equal to $(1/2) \int_{Z_m} \sum_{e \in E_m} c(e, \eta) \left(\pi_e \sqrt{\varphi}\right)^2 d\nu_m$. For every $C > 0$, we define the subset of $\mathcal{P}(Z_m)$ by

$$\mathcal{P}_{m,C} := \left\{ \mu \in \mathcal{P}(Z_m) \ : \ \mu \text{ is } \Gamma/\Gamma_m\text{-invariant and } I_m(\mu) \le C \frac{[\Gamma : \Gamma_m]}{t_m} \right\}.$$

Theorem 3.1 *For every $C > 0$, it holds that*

$$\lim_{i \to \infty} \limsup_{\varepsilon \to 0} \limsup_{m \to \infty} \sup_{\mu \in \mathcal{P}_{m,C}} \mathbb{E}_\mu \left| \overline{f}_{o,i} - \langle f_o \rangle (\overline{\eta}_{o,b(\varepsilon\sqrt{t_m})}) \right| = 0.$$

First, we see how to deduce Theorem 2.1 from Theorem 3.1.

Proof of Theorem 2.1 As we observe, it suffices to prove the required estimate for $\mathbb{P}_m^{eq}$. Recall that $V_{o,m,\varepsilon,i}(\eta) = \sum_{\underline{\sigma} \in \Gamma/\Gamma_m} \left| \overline{f}_{\underline{\sigma} o,i}(\eta) - \langle f_o \rangle \left(\overline{\eta}_{\underline{\sigma} o, b(\varepsilon\sqrt{t_m})} \right) \right|$.

By the Chebychev inequality, for all $a > 0$ and for all $\delta > 0$,

$$\mathbb{P}_m^{eq} \left(\frac{1}{[\Gamma : \Gamma_m]} \int_0^T V_{o,m,\varepsilon,i}(\eta(t)) dt \ge \delta \right) \le \mathbb{E}_m^{eq} \exp \left(a \int_0^T V_{o,m,\varepsilon,i}(\eta(t)) dt - a\delta[\Gamma : \Gamma_m] \right).$$

For all $a > 0$, we consider the operator

$$t_m L_m + a V_{o,m,\varepsilon,i} : L^2(Z_m, \nu_m) \to L^2(Z_m, \nu_m),$$

which is self-adjoint for all $a > 0$ by the definition of L_m. Denote by $\lambda_{o,m,\varepsilon,i}(a)$ the largest eigenvalue of $t_m L_m + a V_{o,m,\varepsilon,i}$. By the Feynmann-Kac formula (e.g., [8][Lemma 7.2, Appendix 1]),

$$\mathbb{E}_m^{eq} \exp \left(a \int_0^T V_{o,m,\varepsilon,i} dt \right) \le \exp T \lambda_{o,m,\varepsilon,i}(a).$$

Therefore it suffices to show that for all $a > 0$,

$$\lim_{i \to \infty} \limsup_{\varepsilon \to 0} \limsup_{m \to \infty} \frac{1}{[\Gamma : \Gamma_m]} \lambda_{o,m,\varepsilon,i}(a) = 0. \tag{3.1}$$

Indeed, by using (3.1), we have that

$$\lim_{i \to \infty} \limsup_{\varepsilon \to 0} \limsup_{m \to \infty} \frac{1}{[\Gamma : \Gamma_m]} \log \mathbb{P}_m^{eq} \left(\frac{1}{[\Gamma : \Gamma_m]} \int_0^T V_{o,m,\varepsilon,i}(\eta(t)) dt \ge \delta \right) \le -a\delta.$$

Taking $a \to \infty$, we obtain Theorem 2.1.

It remains to prove (3.1). By the variational principle, the largest eigenvalue $\lambda_{o,m,\varepsilon,i}(a)$ can be expressed in the following form:

$$\lambda_{o,m,\varepsilon,i}(a) := \sup_{\mu \in \mathcal{P}(Z_m)} \left\{ a \int_{Z_m} V_{o,m,\varepsilon,i} d\mu - t_m I_m(\mu) \right\}.$$

It is enough to consider only the case when $a \int_{Z_m} V_{o,m,\varepsilon,i} d\mu \geq t_m I_m(\mu)$ for some $\mu \in \mathcal{P}(Z_m)$. For $\mu \in \mathcal{P}(Z_m)$, we denote by $\overline{\mu}$ the average of μ by the Γ/Γ_m-action, that is,

$$\overline{\mu} := \frac{1}{[\Gamma : \Gamma_m]} \sum_{\underline{\sigma} \in \Gamma/\Gamma_m} \mu \circ \underline{\sigma}.$$

Here $\overline{\mu}$ is a Γ/Γ_m-invariant probability measure. Now we have that

$$\frac{1}{[\Gamma : \Gamma_m]} \int_{Z_m} V_{o,m,\varepsilon,i} d\mu = \mathbb{E}_{\overline{\mu}} \left| \overline{f}_{o,i} - \langle f_o \rangle (\overline{\eta}_{o,b(\varepsilon\sqrt{t_m})}) \right|.$$

There exists a constant $C(f) > 0$ depending only on f such that

$$V_{o,m,\varepsilon,i} \leq C(f)[\Gamma : \Gamma_m].$$

By the convexity of I_m,

$$I_m(\overline{\mu}) \leq \frac{1}{[\Gamma : \Gamma_m]} \sum_{\underline{\sigma} \in \Gamma/\Gamma_m} I_m(\mu \circ \underline{\sigma}),$$

and since I_m is Γ/Γ_m-invariant, we have that

$$I_m(\overline{\mu}) \leq aC(f) \frac{[\Gamma : \Gamma_m]}{t_m}.$$

When we denote by $\mathcal{P}_{m,aC(f)}$ the set of probability measures μ which is Γ/Γ_m-invariant with $I_m(\mu) \leq aC(f)[\Gamma : \Gamma_m]/t_m$, then

$$\frac{1}{[\Gamma : \Gamma_m]} \lambda_{o,m,\varepsilon,i}(a) \leq a \sup_{\mu \in \mathcal{P}_{m,aC(f)}} \mathbb{E}_{\mu} \left| \overline{f}_{o,i} - \langle f_o \rangle (\overline{\eta}_{o,b(\varepsilon\sqrt{t_m})}) \right|.$$

Hence (3.1) follows from Theorem 3.1. □

Henceforth we often identify a probability measure on Z_m with the one on Z by the periodic extension: Let $\pi_m : V \to V_m$ be the covering map induced by the Γ-action. Define a periodic inclusion $\iota_m : Z_m \to Z$ by $(\iota_m \eta)_z := \eta_{\pi_m(z)}$, $\eta \in Z_m$, $z \in V$. We identify μ on Z_m with its push forward by ι_m, which is a periodic extension of μ on Z. Conversely, we identify a Γ_m-invariant probability measure on Z with the one on Z_m in a natural way.

Theorem 3.1 follows the one-block estimate (Theorem 3.2) and the two-blocks estimate (Theorem 3.3). First, we prove the one-block estimate.

Theorem 3.2 (The one-block estimate) *For every Γ-invariant local function bundle $f : V \times Z \to \mathbb{R}$ and for every $C > 0$, it holds that*

$$\lim_{i\to\infty} \limsup_{m\to\infty} \sup_{\mu\in\mathcal{P}_{m,C}} \mathbb{E}_\mu \left|\overline{f}_{o,i} - \langle f_o\rangle(\overline{\eta}_{o,i})\right| = 0.$$

We discuss a restricted region in X and define the corresponding Dirichlet form. Let $\Lambda = (V_\Lambda, E_\Lambda)$ be a subgraph of X. Define the restricted configuration space by $Z_\Lambda := \{0,1\}^{V_\Lambda}$, and the $(1/2)$-Bernoulli measure ν_Λ on Z_Λ by the direct product of the $(1/2)$-Bernoulli measures on $\{0,1\}$.

In our setting, Γ acts on X vertex transitively, thus $o \in V$ is a fundamental domain in V. We can choose a fundamental domain E^0 in E as the set of those edges; oe or $te = o$. We use the same notation E^0 for the image of E^0 on X_m via the covering map. Define the operator on $L^2(Z_\Lambda, \nu_\Lambda)$ by

$$L_\Lambda^\circ := \frac{1}{2}\sum_{e\in E_\Lambda} \pi_e.$$

For $\mu \in \mathcal{P}(Z)$, we denote by $\mu|_\Lambda$ the restriction of μ on Z_Λ, that is, defined by taking the average outside of Λ. Set $\varphi_\Lambda := d\mu|_\Lambda/d\nu_\Lambda$ the density of $\mu|_\Lambda$. The corresponding Dirichlet form of $\sqrt{\varphi_\Lambda}$ is

$$I_\Lambda^\circ(\mu) := -\int_{Z_\Lambda} \sqrt{\varphi_\Lambda} L_\Lambda^\circ \sqrt{\varphi_\Lambda} d\nu_\Lambda.$$

Proof of Theorem 3.2 Define a subgraph Λ of X by setting $V_\Lambda := B(o, K)$ and $E_\Lambda := \{e \in E \ : \ oe, te \in V_\Lambda\}$. For every $\mu \in \mathcal{P}_{m,C}$, by the convexity of the Dirichlet form and by the (Γ/Γ_m)-invariance of μ and ν_m, we have

$$\begin{aligned}
I_\Lambda^\circ(\mu) &= \frac{1}{2}\sum_{e\in E_\Lambda}\int_{Z_\Lambda} (\pi_e\sqrt{\varphi_\Lambda})^2 d\nu_\Lambda \\
&\le \frac{1}{2}\sum_{e\in E_\Lambda}\int_{Z_m} (\pi_e\sqrt{\varphi})^2 d\nu_m \\
&\le \frac{1}{2}\sum_{\underline{\sigma}\in\Gamma/\Gamma_m, |\sigma|_\Gamma\le K}\ \sum_{e\in\underline{\sigma}E^0}\int_{Z_m} (\pi_e\sqrt{\varphi})^2 d\nu_m \\
&= \frac{1}{2}|B_\Gamma(K)|\sum_{e\in E^0}\int_{Z_m} (\pi_e\sqrt{\varphi})^2 d\nu_m,
\end{aligned}$$

where in the last term $B_\Gamma(K)$ denotes the ball of radius K in Γ about id in Γ by the word norm $|\cdot|_\Gamma$. On the other hand in the same way, we have that

$$I_m(\mu) = [\Gamma : \Gamma_m]\frac{1}{2}\sum_{e\in E^0}\int_{Z_m} c(e,\eta)(\pi_e\sqrt{\varphi})^2 d\nu_m. \tag{3.2}$$

By the uniformly boundedness of $c(e,\eta) \geq c_0 > 0$ (the non-degeneracy of $c(\cdot,\cdot)$), it holds that

$$I^\circ_\Lambda(\mu) \leq \frac{|B_\Gamma(K)|}{c_0[\Gamma : \Gamma_m]} I_m(\mu).$$

Since μ satisfies $I_m(\mu) \leq C[\Gamma : \Gamma_m]/t_m$, we have that $I^\circ_\Lambda(\mu) \leq C_K/t_m \to 0$ as $m \to \infty$, where C_K is a constant depending only on K. The space of probability measures $\mathcal{P}(Z)$ is compact with the weak topology. Thus, every sequence $\{\mu_i\}_{i=1}^\infty$ in $\mathcal{P}(Z)$ has a convergence subsequence. Let $\mathcal{A} \subset \mathcal{P}(Z)$ be the set of all limit points of $\{\mu_i\}_{i=1}^\infty$ in $\mathcal{P}(Z)$. By the above argument, we have that $I^\circ_\Lambda(\mu) = 0$ for every $\mu \in \mathcal{A}$. We obtain that $\mu|_\Lambda(\eta^e) = \mu|_\Lambda(\eta)$ for all $e \in E_\Lambda$ and all $\eta \in Z_\Lambda$ since

$$I^\circ_\Lambda(\mu) = \frac{1}{2}\sum_{e\in E_\Lambda}\sum_{\eta\in Z_\Lambda}(\pi_e\sqrt{\mu|_\Lambda(\eta)})^2 = 0.$$

This holds for every Λ with radius K and implies that random variables $\{\eta_x\}_{x\in V}$ are exchangeable under μ. By the de Finetti theorem, there exists a probability measure λ on $[0,1]$ such that $\mu = \int_0^1 \nu_\rho\lambda(d\rho)$, where ν_ρ is the ρ-Bernoulli measure on Z. Then we have

$$\begin{aligned}\limsup_{m\to\infty}\ \sup_{\mu\in\mathcal{P}_{m,C}}\ \mathbb{E}_\mu\left|\overline{f}_{o,i} - \langle f_o\rangle(\overline{\eta}_{o,i})\right| &\leq \sup_{\mu\in\mathcal{A}}\mathbb{E}_\mu\left|\overline{f}_{o,i} - \langle f_o\rangle(\overline{\eta}_{o,i})\right| \\ &\leq \sup_{\rho\in[0,1]}\mathbb{E}_{\nu_\rho}\left|\overline{f}_{o,i} - \langle f_o\rangle(\overline{\eta}_{o,i})\right|.\end{aligned}$$

Therefore, it is enough to show that $\lim_{i\to\infty}\sup_{\rho\in[0,1]}\mathbb{E}_{\nu_\rho}\left|\overline{f}_{o,i} - \langle f_o\rangle(\overline{\eta}_{o,i})\right| = 0$. Since f is a Γ-invariant local function bundle, there exists $L \geq 0$ such that $f_o : Z \to \mathbb{R}$ depends only on $\{\eta_z : d(o,z) \leq L\}$ and there exists a constant $C(f) > 0$ depending only on f such that

$$\mathbb{E}_{\nu_\rho}\left|\overline{f}_{o,i} - \mathbb{E}_{\nu_\rho}[\overline{f}_{o,i}]\right|^2 \leq C(f)C_L/|F_i| \to 0 \tag{3.3}$$

as $i \to \infty$, where C_L is a constant depending only on L. Here we have that $\langle f_o\rangle(\rho) = \mathbb{E}_{\nu_\rho}[\overline{f}_{o,i}]$ since f is Γ-invariant. Since $\langle f_o\rangle(\rho)$ is a polynomial with respect to ρ, in particular, uniformly continuous on [0, 1], it holds that

$$\sup_{\rho\in[0,1]} \mathbb{E}_{\nu_\rho} \left|\langle f_o\rangle(\overline{\eta}_{o,i}) - \langle f_o\rangle(\rho)\right| \to 0, i \to \infty.$$

By the triangular inequality,

$$\sup_{\rho\in[0,1]} \mathbb{E}_{\nu_\rho} \left|\overline{f}_{o,i} - \langle f_o\rangle(\overline{\eta}_{o,i})\right|$$
$$\leq \sup_{\rho\in[0,1]} \mathbb{E}_{\nu_\rho} \left|\overline{f}_{o,i} - \langle f_o\rangle(\rho)\right| + \sup_{\rho\in[0,1]} \mathbb{E}_{\nu_\rho} \left|\langle f_o\rangle(\rho) - \langle f_o\rangle(\overline{\eta}_{o,i})\right| \to 0, i \to \infty.$$

This concludes that $\lim_{i\to\infty} \limsup_{m\to\infty} \sup_{\mu\in\mathcal{P}_{m,C}} \mathbb{E}_\mu \left|\overline{f}_{o,i} - \langle f_o\rangle(\overline{\eta}_{o,i})\right| = 0$. □

Next, we prove the two-blocks estimate.

Theorem 3.3 (The two-blocks estimate)
For every $C > 0$, it holds that

$$\lim_{i\to\infty} \limsup_{\varepsilon\to 0} \limsup_{L\to\infty} \limsup_{m\to\infty} \sup_{\sigma\in\Gamma s.t. L<|\sigma|\leq\varepsilon\sqrt{t_m}} \sup_{\mu\in\mathcal{P}_{m,C}} \mathbb{E}_\mu \left|\overline{\eta}_{o,i} - \overline{\eta}_{\sigma o,i}\right| = 0.$$

We introduce the following notions: Denote by $\mathcal{P}(Z\times Z)$ the space of probability measures on $Z \times Z$. For $\sigma \in \Gamma$, define $\hat{\sigma} : Z \to Z \times Z$ by $\hat{\sigma}(\eta) := (\eta, \sigma^{-1}\eta)$. For $\mu \in \mathcal{P}(Z)$, denote by $\hat{\sigma}\mu \in \mathcal{P}(Z \times Z)$ the push forward by $\hat{\sigma}$, i.e., $\hat{\sigma}\mu := \mu \circ \hat{\sigma}^{-1}$. We define the subset $\mathcal{A}_{\varepsilon,L}$ in $\mathcal{P}(Z \times Z)$ as the set of all limit points of $\{\hat{\sigma}\mu : L < |\sigma| \leq \varepsilon\sqrt{t_m}, \mu \in \mathcal{P}_{m,C}\}$ as $m \to \infty$, and the subset $\mathcal{A}_\varepsilon$ in $\mathcal{P}(Z \times Z)$ as the set of all limit points of $\mathcal{A}_{\varepsilon,L}$ as $L \to \infty$.

Then we have that:

$$\limsup_{L\to\infty} \limsup_{m\to\infty} \sup_{\sigma\in\Gamma \text{s.t.} L<|\sigma|\leq\varepsilon\sqrt{t_m}} \sup_{\mu\in\mathcal{P}_{m,C}} \int_{(\eta,\eta')\in Z\times Z} \left|\overline{\eta}_{o,i} - \overline{\eta}'_{o,i}\right| (\hat{\sigma}\mu)(d\eta d\eta')$$
$$\leq \sup_{\mu\in\mathcal{A}_\varepsilon} \int_{(\eta,\eta')\in Z\times Z} \left|\overline{\eta}_{o,i} - \overline{\eta}'_{o,i}\right| \mu(d\eta d\eta').$$

As in the proof of the one block estimate, we introduce a subgraph Λ in X, by setting $V_\Lambda := B(o, K)$, $E_\Lambda := \{e \in E : oe, te \in V_\Lambda\}$ and consider the generator L°_Λ on $L^2(Z, \nu)$ by

$$L^\circ_\Lambda = \frac{1}{2} \sum_{e\in E_\Lambda} \pi_e.$$

Then let us define the two generators on $L^2(Z \times Z, \nu \otimes \nu)$ and the corresponding Dirichlet forms. First, we define the generator on $L^2(Z \times Z, \nu \otimes \nu)$ by $L^\circ_\Lambda \otimes 1 + 1 \otimes L^\circ_\Lambda$. For $\mu \in \mathcal{P}(Z \times Z)$, denote by $\mu|_{\Lambda\times\Lambda}$ the restriction of μ on $Z_\Lambda \times Z_\Lambda$, i.e., $\mu|_{\Lambda\times\Lambda}(\eta,\eta') := \mu(\{(\tilde{\eta},\tilde{\eta}') : \tilde{\eta}|_\Lambda = \eta, \tilde{\eta}'|_\Lambda = \eta'\})$ for $(\eta,\eta') \in Z_\Lambda \times Z_\Lambda$. The corresponding Dirichlet form of $\sqrt{\varphi_{\Lambda\times\Lambda}}$, where $\varphi_{\Lambda\times\Lambda} := d\mu|_{\Lambda\times\Lambda}/d\nu_\Lambda \otimes \nu_\Lambda$ is defined by

$$I^\circ_{\Lambda\times\Lambda}(\mu) := -\int_{Z_\Lambda\times Z_\Lambda} \sqrt{\varphi_{\Lambda\times\Lambda}}(L^\circ_\Lambda \otimes 1 + 1 \otimes L^\circ_\Lambda)\sqrt{\varphi_{\Lambda\times\Lambda}} d\nu_\Lambda \otimes \nu_\Lambda.$$

Second, we define the generator on $L^2(Z\times Z, \nu\otimes\nu)$ in the following way: For $(x,y) \in V \times V$ and $(\eta,\eta') \in Z \times Z$, we construct a new configuration $(\eta,\eta')^{(x,y)} \in Z \times Z$ by setting (η'_y, η_x) on (x,y), (η'_x, η_y) on (y,x) and keeping unchanged otherwise. For $F \in L^2(Z\times Z, \nu\otimes\nu)$, $(x,y) \in V \times V$, define $\tilde{\pi}_{x,y}F((\eta,\eta')) = F((\eta,\eta')^{(x,y)}) - F((\eta,\eta'))$. Define the generator $L^\circ_{o,o}$ on $L^2(Z\times Z, \nu\otimes\nu)$ by

$$L^\circ_{o,o} := \tilde{\pi}_{o,o},$$

and the corresponding Dirichlet form of $\sqrt{\varphi_{\Lambda\times\Lambda}}$ by

$$I^{(o,o)}_{\Lambda\times\Lambda}(\mu) := -\int_{Z_\Lambda\times Z_\Lambda} \sqrt{\varphi_{\Lambda\times\Lambda}} L^\circ_{o,o} \sqrt{\varphi_{\Lambda\times\Lambda}} d\nu_\Lambda \otimes \nu_\Lambda.$$

Then we use the following lemma. The proof appears in Lemma 4.2 in [17], so we omit it.

Lemma 3.1 *For all $F \in L^2(Z_m, \nu_m)$ and all $\underline{\sigma} \in \Gamma/\Gamma_m$, it holds that*

$$\int_{Z_m} (\pi_{o,\underline{\sigma}o}F)^2 d\nu_m \le 4d(o,\underline{\sigma}o)^2 \sum_{e\in E^0} \int_{Z_m} (\pi_e F)^2 d\nu_m.$$

For any constant $\tilde{C} > 0$, let

$$\mathcal{A}_{\varepsilon,\tilde{C}} := \{\mu \in \mathcal{P}(Z\times Z) : I^\circ_{\Lambda\times\Lambda}(\mu) = 0, I^{(o,o)}_{\Lambda\times\Lambda}(\mu) \le \tilde{C}\varepsilon^2\}.$$

Then we have the following lemma.

Lemma 3.2 *There exists a constant $\tilde{C} > 0$ such that $\mathcal{A}_\varepsilon \subset A_{\varepsilon,\tilde{C}}$.*

Proof Recall that a subgraph $\Lambda = (V_\Lambda, E_\Lambda)$ of X is defined by $V_\Lambda := B(o,K)$ and $E_\Lambda := \{e \in E : oe, te \in V_\Lambda\}$. Here we regard X_m as a suitable fundamental domain in X by the Γ_m-action. Take large enough L, and m for the diameter of Λ so that for all σ with $L < |\sigma|$, $V_\Lambda \cap \sigma V_\Lambda = \emptyset$, for all σ with $L < |\sigma| \le \varepsilon\sqrt{t_m}$, $V_\Lambda \cap \sigma V_\Lambda \subset X_m$. Notice that $\sqrt{t_m} \le 2 diam X_m$.

For given $(\eta,\eta') \in Z_\Lambda \times Z_\Lambda$, we define $\tilde{\eta} \in Z_{\Lambda\cup\sigma\Lambda}$ by $\tilde{\eta}|_\Lambda = \eta, \sigma^{-1}\tilde{\eta}|_\Lambda = \eta'$. Then $(\hat{\sigma}\mu)|_{\Lambda\times\Lambda}(\eta,\eta') = \mu|_{\Lambda\cup\sigma\Lambda}(\tilde{\eta})$. By using the generator $L^\circ_{\Lambda\cup\sigma\Lambda}$ on $L^2(Z,\nu)$, we have that $I^\circ_{\Lambda\times\Lambda}(\hat{\sigma}\mu) = I^\circ_{\Lambda\cup\sigma\Lambda}(\mu)$. For every $\mu \in \mathcal{P}_{m,C}$ as in the previous section by using (3.2), and by the convexity of the Dirichlet form and the (Γ/Γ_m)-invariance of μ and ν_m, we have that

$$I^{\circ}_{\Lambda\times\Lambda}(\hat{\sigma}\mu) = I^{\circ}_{\Lambda\cup\sigma\Lambda}(\mu) \le \frac{|B_\Gamma(K)|}{2c_o[\Gamma:\Gamma_m]} I_m(\mu) \le C\frac{|B_\Gamma(K)|}{2c_0 t_m} \to 0,$$

as $m \to \infty$. Therefore $I^{\circ}_{\Lambda\times\Lambda}(\mu_L) = 0$ for all $\mu_L \in \mathcal{A}_{\varepsilon,L}$, and $I^{\circ}_{\Lambda\times\Lambda}(\tilde{\mu}) = 0$ for all $\tilde{\mu} \in \mathcal{A}_\varepsilon$ by the continuity of $I^{\circ}_{\Lambda\times\Lambda}$. Furthermore, by the convexity of the Dirichlet form,

$$I^{(o,o)}_{\Lambda\times\Lambda}(\hat{\sigma}\mu) = \frac{1}{2}\int_{Z_{\Lambda\cup\sigma\Lambda}} \left(\pi_{o,\underline{\sigma}o}\sqrt{\frac{d\mu|_{\Lambda\cup\sigma\Lambda}}{d\nu_{\Lambda\cup\sigma\Lambda}}}\right)^2 d\nu_{\Lambda\cup\sigma\Lambda} \le \frac{1}{2}\int_{Z_m} \left(\pi_{o,\underline{\sigma}o}\sqrt{\frac{d\mu}{d\nu_m}}\right)^2 d\nu_m.$$

By Lemma 3.1, the (Γ/Γ_m)-invariance of $\mu \in \mathcal{P}_{m,C}$ and (3.2),

$$\begin{aligned}\int_{Z_m}\left(\pi_{o,\underline{\sigma}o}\sqrt{\frac{d\mu}{d\nu_m}}\right)^2 d\nu_m &\le 4d(o,\underline{\sigma}o)^2 \sum_{e\in E^0}\int_{Z_m}\left(\pi_e\sqrt{\frac{d\mu}{d\nu_m}}\right)^2 d\nu_m \\ &\le 4d(o,\underline{\sigma}o)^2 \frac{2}{c_0[\Gamma:\Gamma_m]} I_m(\mu).\end{aligned}$$

Since $d(o,\underline{\sigma}o) \le |\sigma|_\Gamma$, we have that for $\sigma \in \Gamma$ with $L < |\sigma| \le \varepsilon\sqrt{t_m}$

$$I^{(o,o)}_{\Lambda\times\Lambda}(\hat{\sigma}\mu) \le \frac{1}{2}\cdot 4(\varepsilon\sqrt{t_m})^2 \frac{2}{[\Gamma:\Gamma_m]}\frac{C}{c_0}\frac{[\Gamma:\Gamma_m]}{t_m} = \frac{4C}{c_0}\varepsilon^2.$$

Define $\tilde{C} := 4C/c_0$. By the continuity of $I^{(o,o)}_{\Lambda\times\Lambda}$, we have that $I^{(o,o)}_{\Lambda\times\Lambda}(\tilde{\mu}) \le \tilde{C}\varepsilon^2$ for all $\tilde{\mu} \in \mathcal{A}_\varepsilon$. This implies that $\mathcal{A}_\varepsilon \subset \mathcal{A}_{\varepsilon,\tilde{C}}$. □

Proof of Theorem 3.3 Let $\mathcal{A}_0$ be the set of all limit points of $\mathcal{A}_{\varepsilon,\tilde{C}}$ as $\varepsilon \to 0$. It holds that $I^{\circ}_{\Lambda\times\Lambda}(\tilde{\mu}_0) = 0$ and $I^{(o,o)}_{\Lambda\times\Lambda}(\tilde{\mu}_0) = 0$ for all $\tilde{\mu}_0 \in \mathcal{A}_0$ on each Λ. This shows that the random variables $\{(\eta_x, \eta'_y)\}_{x,y\in V}$ are exchangeable on $Z \times Z$ under $\tilde{\mu}_0$. By the de Finetti theorem, there exists a probability measure λ on $[0, 1]$ such that

$$\tilde{\mu}_0 = \int_{[0,1]} \nu_\rho \otimes \nu_\rho \lambda(d\rho).$$

In the proof of Theorem 3.2 (3.3), we have

$$\lim_{i\to\infty} \sup_{\rho\in[0,1]} \mathbb{E}_{\nu_\rho} \left|\overline{\eta}_{o,i} - \rho\right|^2 = 0.$$

Then

$$\sup_{\tilde{\mu}_0 \in \mathcal{A}_0} \mathbb{E}_{\tilde{\mu}_0} \left| \overline{\eta}_{o,i} - \overline{\eta}'_{o,i} \right| \leq \sup_{\rho \in [0,1]} \mathbb{E}_{\nu_\rho \otimes \nu_\rho} \left| \overline{\eta}_{o,i} - \overline{\eta}'_{o,i} \right| \leq 2 \sup_{\rho \in [0,1]} \mathbb{E}_{\nu_\rho} \left| \overline{\eta}_{o,i} - \rho \right| \to 0,$$

as $i \to \infty$. Here we used the triangular inequality in the last inequality. By Lemma 3.2, we have that $\mathcal{A}_\varepsilon \subset \mathcal{A}_{\varepsilon,\tilde{C}}$ for some $\tilde{C} > 0$,

$$\begin{aligned}
&\limsup_{\varepsilon\to 0} \limsup_{L\to\infty} \limsup_{m\to\infty} \sup_{\sigma\in\Gamma \text{s.t.} L<|\sigma|\leq\varepsilon\sqrt{t_m}} \sup_{\mu\in\mathcal{P}_{m,C}} \mathbb{E}_\mu \left| \overline{\eta}_{o,i} - \overline{\eta}_{\underline{\sigma o},i} \right| \\
&\leq \limsup_{\varepsilon\to 0} \sup_{\tilde{\mu}\in\mathcal{A}_\varepsilon} \mathbb{E}_{\tilde{\mu}} \left| \overline{\eta}_{o,i} - \overline{\eta}'_{o,i} \right| \\
&\leq \limsup_{\varepsilon\to 0} \sup_{\tilde{\mu}\in\mathcal{A}_{\varepsilon,\tilde{C}}} \mathbb{E}_{\tilde{\mu}} \left| \overline{\eta}_{o,i} - \overline{\eta}'_{o,i} \right| \\
&\leq \sup_{\tilde{\mu}_0\in\mathcal{A}_0} \mathbb{E}_{\tilde{\mu}_0} \left| \overline{\eta}_{o,i} - \overline{\eta}'_{o,i} \right| \to 0,
\end{aligned}$$

as $i \to \infty$. This proves the theorem. □

Proof of Theorem 3.1 For $\eta \in Z$, the following uniform estimates hold:

$$\left| \overline{\eta}_{o,b(\varepsilon\sqrt{t_m})} - \frac{1}{\left|F_{b(\varepsilon\sqrt{t_m})}\right|} \sum_{\sigma\in F_{b(\varepsilon\sqrt{t_m})}} \overline{\eta}_{\sigma o,i} \right| \leq C(F_i) \frac{\left|\partial_S F_{b(\varepsilon\sqrt{t_m})}\right|}{\left|F_{b(\varepsilon\sqrt{t_m})}\right|}, \tag{3.4}$$

where $C(F_i)$ is a constant depending only on F_i, and

$$\left| \frac{1}{\left|F_{b(\varepsilon\sqrt{t_m})}\right|} \sum_{\sigma\in F_{b(\varepsilon\sqrt{t_m})}} \overline{\eta}_{\sigma o,i} - \frac{1}{\left|F_{b(\varepsilon\sqrt{t_m})}\right|} \sum_{\sigma\in F_{b(\varepsilon\sqrt{t_m})}\setminus B_\Gamma(L)} \overline{\eta}_{\sigma o,i} \right| \leq \frac{|B_\Gamma(L)|}{\left|F_{b(\varepsilon\sqrt{t_m})}\right|}. \tag{3.5}$$

By the triangular inequality, (3.4) and (3.5), we have that

$$\left| \overline{\eta}_{o,b(\varepsilon\sqrt{t_m})} - \frac{1}{\left|F_{b(\varepsilon\sqrt{t_m})}\setminus B_\Gamma(L)\right|} \sum_{\sigma\in F_{b(\varepsilon\sqrt{t_m})}\setminus B_\Gamma(L)} \overline{\eta}_{\sigma o,i} \right| \leq C(F_i) \frac{\left|\partial_S F_{b(\varepsilon\sqrt{t_m})}\right|}{\left|F_{b(\varepsilon\sqrt{t_m})}\right|} + 2\frac{|B_\Gamma(L)|}{\left|F_{b(\varepsilon\sqrt{t_m})}\right|},$$

where the last terms go to 0 as $m \to \infty$, since a Følner sequence satisfies that $|\partial_S F_{b(\varepsilon\sqrt{t_m})}|/|F_{b(\varepsilon\sqrt{t_m})}| \to 0$ as $m \to \infty$.

For every $\mu \in \mathcal{P}_{m,C}$, we have that

$$\mathbb{E}_\mu \left| \overline{\eta}_{o,i} - \overline{\eta}_{o,b(\varepsilon\sqrt{t_m})} \right| \leq \mathbb{E}_\mu \left| \overline{\eta}_{o,i} - \frac{1}{\left|F_{b(\varepsilon\sqrt{t_m})}\setminus B_\Gamma(L)\right|} \sum_{\sigma\in F_{b(\varepsilon\sqrt{t_m})}\setminus B_\Gamma(L)} \overline{\eta}_{\sigma o,i} \right| + o_m,$$

where $o_m \to 0$ as $m \to \infty$. The last term is bounded by

$$\frac{1}{\left|F_{b(\varepsilon\sqrt{t_m})} \setminus B_\Gamma(L)\right|} \sum_{\sigma \in F_{b(\varepsilon\sqrt{t_m})} \setminus B_\Gamma(L)} \mathbb{E}_\mu \left|\overline{\eta}_{o,i} - \overline{\eta}_{\sigma o,i}\right| \leq \sup_{\sigma \in F_{b(\varepsilon\sqrt{t_m})} \setminus B_\Gamma(L)} \mathbb{E}_\mu \left|\overline{\eta}_{o,i} - \overline{\eta}_{\sigma o,i}\right|$$

$$\leq \sup_{\sigma \in \Gamma \text{s.t.} L < |\sigma|_\Gamma \leq \varepsilon\sqrt{t_m}} \sup_{\mu \in \mathcal{P}_{m,C}} \mathbb{E}_\mu \left|\overline{\eta}_{o,i} - \overline{\eta}_{\sigma o,i}\right|.$$

By Theorem 3.3, it holds that

$$\lim_{i\to\infty} \limsup_{\varepsilon\to 0} \limsup_{m\to\infty} \sup_{\mu \in \mathcal{P}_{m,C}} \mathbb{E}_\mu \left|\overline{\eta}_{o,i} - \overline{\eta}_{o,b(\varepsilon\sqrt{t_m})}\right| = 0.$$

Since for every Γ-periodic local function bundle $f : V \times Z \to \mathbb{R}$, the function $\langle f_o \rangle(\cdot)$ is uniformly continuous on $[0, 1]$,

$$\lim_{i\to\infty} \limsup_{\varepsilon\to 0} \limsup_{m\to\infty} \sup_{\mu \in \mathcal{P}_{m,C}} \mathbb{E}_\mu \left|\langle f_o \rangle(\overline{\eta}_{o,i}) - \langle f_o \rangle(\overline{\eta}_{o,b(\varepsilon\sqrt{t_m})})\right| = 0.$$

By Theorem 3.2 and the triangular inequality, we conclude that

$$\lim_{i\to\infty} \limsup_{\varepsilon\to 0} \limsup_{m\to\infty} \sup_{\mu \in \mathcal{P}_{m,C}} \mathbb{E}_\mu \left|\overline{f}_{o,i} - \langle f_o \rangle(\overline{\eta}_{o,b(\varepsilon\sqrt{t_m})})\right| = 0,$$

and this completes the proof.

Acknowledgements The author thanks Hiroshi Kawabi for helpful comments and encouragement, Motoko Kotani for support during this work and an anonymous referee for useful comments to improve this note.

References

1. Bartholdi, L., Grigorchuk, R., Nekrashevych, V.: From fractal groups to fractal sets. In: Fractals in Graz 2001. Trends Mathematics, pp. 25–118. Birkhäuser, Basel (2003)
2. Bartholdi, L., Virág, B.: Amenability via random walks. Duke Math. J. **130**(1), 39–56 (2005)
3. Ceccherini-Silberstein, T., Coornaert, M.: Cellular Automata and Groups. Springer Monographs in Mathematics. Springer-Verlag, Berlin (2010)
4. Funaki, T., Handa, K., Uchiyama, K.: Hydrodynamic limit of one-dimensional exclusion processes with speed change. Ann. Probab. **19**(1), 245–265 (1991)
5. Guo, M.Z., Papanicolaou, G.C., Varadhan, S.R.S.: Nonlinear diffusion limit for a system with nearest neighbor interactions. Comm. Math. Phys. **118**(1), 31–59 (1988)
6. Jara, M.: Hydrodynamic limit for a zero-range process in the Sierpinski gasket. Comm. Math. Phys. **288**(2), 773–797 (2009)
7. Kaimanovich, V.A.: Random walks on Sierpiński graphs: hyperbolicity and stochastic homogenization. In: Fractals in Graz 2001. Trends Mathematics, pp. 145–183. Birkhäuser, Basel (2003)

8. Kipnis, C., Landim, C.: Scaling Limits of Interacting Particle Systems. Grundlehren der Mathematischen Wissenschaften, vol. 320. Springer, Berlin (1999)
9. Kipnis, C., Landim, C., Olla, S.: Hydrodynamical limit for a nongradient system: the generalized symmetric exclusion process. Comm. Pure Appl. Math. **47**(11), 1475–1545 (1994)
10. Kipnis, C., Olla, S., Varadhan, S.R.S.: Hydrodynamics and large deviation for simple exclusion processes. Comm. Pure Appl. Math. **42**(2), 115–137 (1989)
11. Kotani, M., Sunada, T.: Albanese maps and off diagonal long time asymptotics for the heat kernel. Comm. Math. Phys. **209**(3), 633–670 (2000)
12. Nekrashevych, V.: Self-Similar Groups. Mathematical Surveys and Monographs, vol. 117. American Mathematical Society, Providence (2005)
13. Olla, S., Sasada, M.: Macroscopic energy diffusion for a chain of anharmonic oscillators. Probab. Theory Relat. Fields **157**(3–4), 721–775 (2013)
14. Sasada, M.: Spectral gap for stochastic energy exchange model with nonuniformly positive rate function. Ann. Probab. **43**(4), 1663–1711 (2015)
15. Sasada, M., Kametani, Y.: A new technique for the computation of central limit theorem variances for exclusion processes and its application. Talk at Research Institute for Mathematical Sciences, Kyoto University (2015)
16. Spohn, H.: Large Scale Dynamics of Interacting Particles. Texts and Monograph in Physics. Springer-Verlag, Heidelberg (1991)
17. Tanaka, R.: Hydrodynamic limit for weakly asymmetric simple exclusion processes in crystal lattices. Comm. Math. Phys. **315**(3), 603–641 (2012)

Unboundedness of Some Solutions to Isentropic Model Equations for the One Dimensional Periodic Motions of a Compressible Self-Gravitating Viscous Fluid

Masahiro Sawada and Yoshitaka Yamamoto

Abstract We consider the initial value problem for the one dimensional system of equations of a compressible viscous fluid driven by a bounded periodic self-gravitation of the fluid. For the initial-boundary value problem on a finite interval with a general bounded forcing term Matsumura and Nishida (Periodic solutions of a viscous gas equation. In: Mimura M, Nishida T (eds) Recent Topics in nonlinear PDE IV. Lecture notes in numerical and applied analysis, vol 10. North-Holland, Amsterdam, pp 49–82, 1989) showed that the isothermal system has a unique global-in-time solution for any initial data, being bounded as well. For the isentropic system Matsumura and Yanagi (Commun Math Phys 175:259–274, 1996) observed that given any initial data and any bounded forcing term, if the adiabatic constant is chosen suitably close to one, then the boundedness of the solution remains true. Without such a choice of the adiabatic constant, however, it is unknown whether the solution is bounded or not. In this paper we focus on the structure of stationary solutions to the isentropic self-gravitational system and prove that a certain bounded stationary solution is lost when the average of the specific volume reaches a critical value. We then show that there dose exist an unbounded solution for the initial value problem when the average exceeds the critical value. We also present a sufficient condition for the unboundedness in terms of the initial values of an energy form.

Keywords Unbounded solutions • Isentropic model equations • Self-gravitation • Stationary solutions • Energy form

---

M. Sawada (✉) • Y. Yamamoto
Graduate School of Information Science and Technology, Osaka University, 1-5 Yamadaoka, Suita 565-0871, Japan
e-mail: masahiro.sawada@ist.osaka-u.ac.jp; yamamoto@ist.osaka-u.ac.jp

Y. Nishiura, M. Kotani (eds.), *Mathematical Challenges in a New Phase of Materials Science*, Springer Proceedings in Mathematics & Statistics 166,
DOI 10.1007/978-4-431-56104-0_7

1 Introduction

The barotropic model for the one dimensional motion of a compressible viscous fluid is described by the following system of equations in the Lagrangian mass coordinates:

$$\begin{cases} \partial_t v - \partial_x u = 0, \\ \partial_t u + \partial_x(p(v)) - \mu \partial_x \left(\dfrac{\partial_x u}{v} \right) = \mathscr{G}, \end{cases} \tag{1}$$

where specific volume v and velocity u of the fluid are unknown functions of the time and space variables $t \geq 0$ and $x \in \mathbf{R}$, pressure $p(v)$ a function of $v > 0$, and $\mu > 0$ the viscosity constant. The second member $\mathscr{G}$ represents an external force specified below. We are concerned with an L-periodic flow, i.e.,

$$v(t, x+L) = v(t, x), \quad u(t, x+L) = u(t, x) \tag{2}$$

driven by a self-gravitation of the fluid:

$$\mathscr{G}(t, x) = -\frac{4\pi G}{\bar{v}} \partial_x \int_0^L K_L(x, y)(v(t, y) - \bar{v})dy, \tag{3}$$

where $K_L(x, y)$ is the Green kernel of the operator $-\partial_x^2$ on the L-periodic functions with average 0:

$$K_L(x, y) = \sum_{n=1}^{\infty} \frac{L}{2\pi^2 n^2} \cos \frac{2\pi n}{L}(x - y),$$

or

$$K_L(x, y) = -\frac{|x-y|}{2} + \frac{(x-y)^2}{2L} + \frac{L}{12}, \quad 0 \leq x, y \leq L, \tag{4}$$

$\bar{v}$ the average of the specific volume:

$$\bar{v} = \frac{1}{L} \int_0^L v(t, x)dx,$$

and $G > 0$ the gravitational constant. This is the representation in the Lagrangian mass coordinates of a field of force often adopted in the classical theory of gravitational instability for an infinite homogeneous fluid. See Weinberg [7], Chapter 15. The astrophysical theory takes into account only the part of gravitational field corresponding to the disturbance in the homogeneous fluid so that the field is consistent with static equilibria of the fluid. The modification contradicts the usual

Poisson equation of gravity, however, for the periodic mass distribution concerned, the field can be derived from Newton's law of gravitation by a summation technique of oscillatory integral. Notice that the average $\bar{v}$ as well as that of u is a constant of motion in view of (1), and therefore the forcing term $\mathscr{G}$ is a bounded function of the variables t and x.

Concerning the initial or initial-boundary value problem for (1) with a prescribed forcing term $\mathscr{G}$ several results refer to bounded solutions. When $\mathscr{G} \equiv 0$ under the assumption $p'(v) < 0$, Kanel' [3] gave an a priori estimate of solutions to the system on the whole line and proved that the solutions with appropriately smooth initial data are global and bounded, i.e.,

$$\sup_{t,x} v(t,x) < \infty.$$

For the system on a finite interval with a general bounded forcing term $\mathscr{G}$ similar results are known, especially in case $p(v)$ takes the form $av^{-\gamma}$ with constants $a > 0$ and $\gamma \geq 1$. In the isothermal case, i.e., $\gamma = 1$, Matsumura and Nishida [5] presented an estimate showing the boundedness of solutions with smooth initial data. This result was extended, in some sense, to the isentropic case, i.e., $\gamma > 1$, by Matsumura and Yanagi [6], but the result requires smallness of $\gamma - 1$ depending on the data.

As for the self-gravitational system the argument of Matsumura and Nishida still works in the isothermal case. Indeed, the boundedness of solutions follows merely from that of the self-gravitational term. In the isentropic case, however, the situation is complicated just because of the non-local character of the term. In spite of rough estimates of the integral term it may actually be small enough to allow us to follow an argument as in Matsumura and Yanagi. If that is not the case the boundedness is left in doubt and may be false. How we can distinguish the cases with prescribed initial data remains open.

In this paper we restrict ourselves to the isentropic case with $1 < \gamma < 2$ and focus on finding out initial conditions for unbounded solutions. Generally speaking, it is a hard task to show the boundless growth of a solution directly on the basis of some a priori bounds of the solution. We prefer to induce initial conditions contradictory to the boundedness of a solution from preliminary studies on the asymptotics of the bounded solutions in general. To be more specific, we first show that the ω-limit set of the orbit associated with a bounded solution constitutes a part of the stationary solutions to the system. We next study the structure and stability of stationary solutions from the viewpoint of the global bifurcation in the stationary problem. We then observe the behavior of an energy form mentioned below near the stationary solutions, and finally conclude that any solution with appropriate average $\bar{v}$ and initial value of the energy form is necessarily unbounded.

The energy form that we introduce for a state (v, u) is given by

$$\mathscr{E}(v,u) = \int_0^L \frac{1}{2} u(x)^2 dx + \mathscr{E}(v) \tag{5}$$

with

$$\mathscr{E}(v) = \int_0^L \int_{\bar{v}}^{v(x)} (p(\bar{v}) - p(\xi))d\xi dx - \frac{2\pi G}{\bar{v}} \int_0^L \int_0^L K_L(x,y)(v(x) - \bar{v})(v(y) - \bar{v})dxdy. \tag{6}$$

Except for the gravitational term the form has already been introduced for the general barotropic model system to estimate bounds of solutions. The outstanding feature of the form is its decrease and boundedness along the orbit of a solution. See **Lemma 1** below. This enables us to successfully incorporate the gravitational field into the analysis of spectral as well as dynamical properties of the system.

The paper is organized as follows. In Sect. 2, after the mention of the global solvability of the initial value problem, we present the main results of the paper in two theorems. One is concerned with the structure of the whole solutions to the stationary problem and the other with an initial condition for unbounded solutions to the initial value problem. In Sect. 3 we observe the asymptotics of bounded solutions to the initial value problem. The results of this section need to be interpreted in comparison with the behavior of the energy form near the stationary solutions. For this purpose Sect. 4 is devoted to the study of the stationary problem with emphasis on certain linear stability of stationary solutions associated with the decreasing property of the energy form. Finally in Sect. 5, combining the structure and stability of stationary solutions with the behavior of the energy form, we present a condition for unbounded solutions in terms of the initial values of the solutions, especially those of the energy form.

To avoid lengthy arguments some of the results of the paper are presented only with the outlines of proofs. The detail will be given in a forthcoming paper.

2 Notation and Main Results

We begin with the notation and a brief comment on the global solvability of the initial value problem for (1), (2), and (3). For a nonnegative integer m and a positive number L let C^m be the space of m times continuously differentiable periodic real-valued functions on $\mathbf{R}$ with period L, and H^m the Sobolev space of locally square integrable L-periodic real-valued functions on $\mathbf{R}$ equipped with scalar product

$$(h_1, h_2)_{H^m} = \sum_{j=0}^{m} \int_0^L \partial_x^j h_1(x) \partial_x^j h_2(x) dx$$

and norm $\|h\|_{H^m} = \sqrt{(h,h)_{H^m}}$. We write $H^0 = L^2$ as usual. Let s be a nonnegative integer and X a Banach space with norm $\|\cdot\|$. The space of s-times continuously differentiable functions on $[0,\infty)$ with values in X is denoted by $C^s([0,\infty);X)$. $H^s_{\text{loc}}(0,\infty;X)$ denotes the space of X-valued strongly measurable functions on

$[0, \infty)$ whose distributional derivatives up to order s are locally square integrable, i.e.,

$$\int_0^T \|\partial_t^k u(t)\|^2 dt < \infty \quad \text{for any } k = 0, \ldots, s \text{ and } T > 0.$$

As concerns the isothermal or isentropic system with a general bounded forcing term it is well-known that the initial-boundary value problem on a finite interval with solid boundary condition has a unique global solution. This follows from the existence of local-in-time solutions and a priori estimates of H^1-norms of the solutions together with the lower bound of v. The method applies also to the initial value problem for (1), (2), and (3) with $p(v) = av^{-\gamma}$ where $a > 0$ and $\gamma \geq 1$ are arbitrary constants, giving a unique global solution (v, u) for any initial value $(v_0, u_0) \in H^1 \times H^1$ with $v_0 > 0$ as follows:

$$\begin{cases} v \in C^1([0,\infty); L^2) \cap C^0([0,\infty); H^1), \quad v(t,\cdot) > 0, \\ u \in H^1_{\text{loc}}(0,\infty; L^2) \cap L^2_{\text{loc}}(0,\infty; H^2). \end{cases} \tag{7}$$

Of interest are unbounded solutions from this class. In what follows we assume that the average of u vanishes, taking $(v, u - \bar{u})$ as new unknown functions.

We present an initial condition for unbounded solutions in connection with the structure of the stationary solutions to (1), (2), and (3). For this purpose let us introduce the following manifold in $H^1 \times H^1$ parametrized by a positive number V:

$$M_V = \{(v, u) \in H^1 \times H^1;\ v > 0,\ \bar{v} = V,\ \bar{u} = 0\}.$$

Clearly, the trivial solution $(V, 0)$ lies in M_V. We further introduce a function I_1 (resp. I_γ, $\gamma > 1$) on the interval $(0, \infty)$(resp. $(0, (\gamma - 1)^{-1/2})$) expressed as

$$I_\gamma(\theta) = \theta \int_0^1 \frac{1}{\sqrt{1-y}} \frac{1}{f_+(F_+{}^{-1}(\theta^2 y))} dy + \theta \int_0^1 \frac{1}{\sqrt{1-y}} \frac{1}{f_-(F_-{}^{-1}(\theta^2 y))} dy, \tag{8}$$

where the functions $f_+(r)$, $F_+(r)$ on $r \geq 0$, and $f_-(r)$, $F_-(r)$ on $0 \leq r < 1$ are given by

$$f_+(r) = 1 - (1 + r)^{-1/\gamma}, \qquad F_+(r) = \int_0^r f_+(s)\, ds,$$

$$f_-(r) = -\{1 - (1 - r)^{-1/\gamma}\}, \quad F_-(r) = \int_0^r f_-(s)\, ds.$$

As shown below in Sect. 3, I_γ is a monotone increasing function with $I_\gamma(\theta) > \sqrt{2\gamma}\pi$ provided that $1 \leq \gamma < 2$. Moreover, $I_\gamma(\theta)$, $\gamma > 1$, has a finite limit as $\theta \to (\gamma - 1)^{-1/2} - 0$. Notice that a non-trivial stationary solution, if exists, has the least period L/j for some positive integer j.

Theorem 1 *Assume* $1 \leq \gamma < 2$. *For* $V > 0$ *let* $k_{\min}$ *and* $k_{\max}$, *respectively, be the smallest and the largest integers* j *satisfying*

$$\left(\frac{a\gamma\pi}{GV^\gamma}\right)^{1/2} < \frac{L}{j} < \begin{cases} \infty, & \gamma = 1, \\ \dfrac{I_\gamma((\gamma-1)^{-1/2}-0)}{\sqrt{2\gamma\pi}} \left(\dfrac{a\gamma\pi}{GV^\gamma}\right)^{1/2}, & 1 < \gamma < 2. \end{cases} \tag{9}$$

Then, for $j = k_{\min}, \ldots, k_{\max}$ *there exists on* M_V *a stationary solution of (1), (2), and (3) with least period* L/j. *The whole stationary solutions lying in* M_V *except for the trivial one are given by* $(\tilde{v}^{(j)}(\cdot - \alpha), 0)$, $0 \leq \alpha < L/j$, $j = k_{\min}, \ldots, k_{\max}$, *where* $(\tilde{v}^{(j)}, 0)$ *is one of the stationary solutions with least period* L/j.

Remark 1 Whenever $V \leq \left(\frac{a\gamma\pi}{GL^2}\right)^{1/\gamma}$, no integer satisfies the condition (9), and hence the stationary problem admits on M_V only the trivial solution. When $V > \left(\frac{a\gamma\pi}{GL^2}\right)^{1/\gamma}$, it depends on the value of γ whether the stationary problem possesses on M_V a solution other than the trivial solution or not. As easily seen, (9) holds with $j = 1$ for $\gamma = 1$. This is true also for $1 < \gamma < 2$ provided that $V < \left(\frac{aI_\gamma((\gamma-1)^{-1/2}-0)^2}{2\pi GL^2}\right)^{1/\gamma}$. In order to describe the situation that takes place when $V \geq \left(\frac{aI_\gamma((\gamma-1)^{-1/2}-0)^2}{2\pi GL^2}\right)^{1/\gamma}$ we need to consult the value of $\frac{I_\gamma((\gamma-1)^{-1/2}-0)}{\sqrt{2\gamma\pi}}$. In fact, if the value exceeds 2, we can choose an integer j subject to (9) for every V. If not, such a choice is impossible for V from a finite number of intervals including $\left(\frac{aI_\gamma((\gamma-1)^{-1/2}-0)^2}{2\pi GL^2}\right)^{1/\gamma} \leq V \leq \left(\frac{4a\gamma\pi}{GL^2}\right)^{1/\gamma}$, and for such a V the trivial solution is the unique stationary solution on M_V. By elementary calculus we can prove that $\lim_{\gamma\to 1+0} I_\gamma((\gamma-1)^{-1/2}-0) = \infty$ and $\lim_{\gamma\to 2-0} I_\gamma((\gamma-1)^{-1/2}-0) = 2\pi$. This implies that the former is true for γ close to 1 and the latter for γ close to 2.

Recall the energy form (5) and (6). In case of $\gamma > 1$ we have

$$\mathscr{E}(v) = \int_0^L a\left(\frac{v(x)-\bar{v}}{\bar{v}^\gamma} - \frac{v(x)^{1-\gamma}-\bar{v}^{1-\gamma}}{1-\gamma}\right)$$

$$\times\, dx - \frac{2\pi G}{\bar{v}} \int_0^L \int_0^L K_L(x,y)(v(x)-\bar{v})(v(y)-\bar{v})\, dxdy. \tag{10}$$

By L-periodicity of v the value of $\mathscr{E}(v)$ is unchanged by replacing v with $v(\cdot - \alpha)$ for any $\alpha \in \mathbf{R}$. We are now ready to state the main theorem.

Theorem 2 *Assume* $1 < \gamma < 2$. *Let* $V \geq \left(\frac{a l_\gamma ((\gamma-1)^{-1/2}-0)^2}{2\pi G L^2}\right)^{1/\gamma}$ *and* $\tilde{v}^{(k_{\min})}$ *be as in Theorem 1.*

(i) *The subset of* $H^1 \times H^1$ *given by*

$$A_V = \left\{ (v, u) \in M_V \,\middle|\, \mathscr{E}(v, u) < \begin{cases} \mathscr{E}(\tilde{v}^{(k_{\min})}), & \text{if integers } j \text{ with (9) exist,} \\ 0, & \text{otherwise} \end{cases} \right\}$$

is nonempty.

(ii) *Any solution of (1), (2), and (3) with initial value from* A_V *is unbounded, i.e.,*

$$\sup_{t,x} v(t, x) = \infty.$$

As suggested by the statement of the theorem and actually shown below in **Proposition 3**, $\mathscr{E}(\tilde{v}^{(k_{\min})})$ if it makes sense or else $\mathscr{E}(V) = 0$ is minimal amongst the values of the energy form evaluated at the stationary solutions on M_V. The theorem claims that there exists a state with value of the energy form smaller than the minimal one provided that $V \geq \left(\frac{a l_\gamma ((\gamma-1)^{-1/2}-0)^2}{2\pi G L^2}\right)^{1/\gamma}$, and that the solution starting from such a state not only keeps away from any stationary solution due to the decrease of the energy form, but also develops boundlessly in amplitude in infinite time.

3 Large Time Behavior of Bounded Solutions

In what follows we consider solutions of (1), (2), and (3) to be subject to (7) unless otherwise stated. First of all, we present a fundamental result on the behavior of the energy form (5), and (6) along the orbits of solutions. This result holds true for any $\gamma \geq 1$ regardless of the boundedness of solutions.

Lemma 1 *For a solution* (v, u) *of (1), (2), and (3) put*

$$E(t) = \mathscr{E}(v(t, \cdot), u(t, \cdot)), \quad t \geq 0. \tag{11}$$

Then we have

$$\frac{dE}{dt}(t) = -\mu \int_0^L \frac{\partial_x u(t, x)^2}{v(t, x)} dx \leq 0, \quad \inf_t E(t) = E(\infty) > -\infty. \tag{12}$$

The formula of dE/dt is derived from (1), (2), and (3) by taking the derivative of E directly. By the symmetry of the integral kernel K_L the derivative turns out to be

$$\int_0^L \Big\{ u(t,x)\partial_t u(t,x) + (p(\bar{v}) - p(v(t,x)))\partial_t v(t,x) \\ - \frac{4\pi G}{\bar{v}} \int_0^L K_L(x,y)(v(t,y) - \bar{v})dy \partial_t v(t,x) \Big\} dx.$$

After substituting $\partial_x u$ for $\partial_t v$, by integration by parts we see that

$$\frac{dE}{dt}(t) = \int_0^L u(t,x) \Big\{ \partial_t u(t,x) + \partial_x(p(v(t,x))) + \frac{4\pi G}{\bar{v}} \partial_x \int_0^L K_L(x,y)(v(t,y) - \bar{v})dy \Big\} dx.$$

With the use of the second equation of (1) and again by integration by parts we obtain the equality as desired. The boundedness of E from below is a consequence of $\int_{\bar{v}}^{v} (p(\bar{v}) - p(\xi))d\xi \geq 0$ and the boundedness of the kernel K_L.

An immediate consequence of (12) is that

$$\mu \int_0^\infty \left(\int_0^L \frac{\partial_x u(t,x)^2}{v(t,x)} dx \right) dt = E(0) - E(\infty) < \infty, \tag{13}$$

which plays an important role in the subsequent study on the large time behavior of bounded solutions.

In the isothermal case $\gamma = 1$ the orbit associated with a solution of (1), (2), and (3) is bounded in $H^1 \times H^1$ with $\inf_{t,x} v(t,x) > 0$. This follows from the same argument as in Matsumura and Nishida [5] with the observation that the forcing term (3) is a bounded function of the variables t and x. Obviously, the boundedness of the orbit implies that of the solution in the sense that $\sup_{t,x} v(t,x) < \infty$. The following lemma claims that the converse itself holds true in the isentropic case with $1 < \gamma \leq 2$.

Lemma 2 *Assume* $1 < \gamma \leq 2$. *For a solution* (v,u) *of (1), (2), and (3) with* $\bar{u} = 0$, *if it is bounded, i.e.,* $\sup_{t,x} v(t,x) < \infty$, *then we have*

$$\sup_t \|v(t,\cdot)\|_{H^1} < \infty, \quad \sup_t \|u(t,\cdot)\|_{H^1} < \infty, \quad \inf_{t,x} v(t,x) > 0. \tag{14}$$

We show (14) by energy estimates of the solution. Regarding the forcing term (3) as a bounded function of the variables t and x, we partly follow the argument of [5] with the use of the equality

$$\frac{d}{dt} \int_0^L \left\{ \frac{1}{2} u(t,x)^2 + a \left(\frac{v(t,x) - \bar{v}}{\bar{v}^\gamma} - \frac{v(t,x)^{1-\gamma} - \bar{v}^{1-\gamma}}{1-\gamma} \right) \right\} dx \\ = -\mu \int_0^L \frac{\partial_x u(t,x)^2}{v(t,x)} dx + \int_0^L \mathscr{G}(t,x)u(t,x)dx \tag{15}$$

combined with another one

$$\frac{d}{dt}\int_0^L \left(\frac{\mu}{2}\frac{\partial_x v(t,x)^2}{v(t,x)^2} - u(t,x)\frac{\partial_x v(t,x)}{v(t,x)}\right) dx$$
$$= -a\gamma \int_0^L \frac{\partial_x v(t,x)^2}{v(t,x)^{\gamma+2}} dx + \int_0^L \frac{\partial_x u(t,x)^2}{v(t,x)} dx - \int_0^L \mathcal{G}(t,x)\frac{\partial_x v(t,x)}{v(t,x)} dx. \quad (16)$$

As in [5], we show that the quantity

$$\int_0^L \left\{ \frac{1}{2}u(t,x)^2 - \frac{\mu}{2}u(t,x)\frac{\partial_x v(t,x)}{v(t,x)} + \frac{\mu^2}{4}\frac{\partial_x v(t,x)^2}{v(t,x)^2} + a\left(\frac{v(t,x)-\bar{v}}{\bar{v}^\gamma} - \frac{v(t,x)^{1-\gamma}-\bar{v}^{1-\gamma}}{1-\gamma}\right)\right\} dx \quad (17)$$

is bounded based on a differential inequality derived from the combination of the form (15)+$(\mu/2)\times$(16). For this purpose an upper bound of (17) in terms of $\int_0^L \frac{\partial_x u(t,x)^2}{v(t,x)} dx$ and $\int_0^L \frac{\partial_x v(t,x)^2}{v(t,x)^{\gamma+2}} dx$ is required. Though u obeys the L-periodic condition instead of the solid boundary condition supposed in [5], the inequality

$$\int_0^L u(t,x)^2 dx \le L^2 \bar{v} \int_0^L \frac{\partial_x u(t,x)^2}{v(t,x)} dx \quad (18)$$

is still available in view of $\bar{u} = 0$. The major difference from [5] is that we estimate bounds of the integrals $\int_0^L \frac{\partial_x v(t,x)^2}{v(t,x)^2} dx$ and $\int_0^L \frac{v(t,x)^{1-\gamma}-\bar{v}^{1-\gamma}}{1-\gamma} dx$ in terms of $\int_0^L \frac{\partial_x v(t,x)^2}{v(t,x)^{\gamma+2}} dx$ with the additional hypothesis that v is bounded, i.e., $\sup_{t,x} v(t,x) < \infty$ under the assumption $1 < \gamma \le 2$. For example, for every $t \ge 0$ choosing $x_t \in [0,L)$ so that $v(t,x_t) = \bar{v}$ holds, by Schwarz' lemma we have

$$\left|\frac{v(t,x)^{1-\gamma}-\bar{v}^{1-\gamma}}{1-\gamma}\right| = \left|\int_{x_t}^x \frac{\partial_y v(t,y)}{v(t,y)^\gamma} dy\right|$$
$$\le \left(\int_0^L v(t,y)^{2-\gamma} dy\right)^{1/2} \left(\int_0^L \frac{\partial_y v(t,y)^2}{v(t,y)^{\gamma+2}} dy\right)^{1/2}$$
$$\le L^{1/2}\left(\sup_{t,y} v(t,y)\right)^{(2-\gamma)/2} \left(\int_0^L \frac{\partial_y v(t,y)^2}{v(t,y)^{\gamma+2}} dy\right)^{1/2}$$

for $x \in [0,L]$, and hence

$$\left|\int_0^L \frac{v(t,x)^{1-\gamma}-\bar{v}^{1-\gamma}}{1-\gamma} dx\right| \le L^{3/2}\left(\sup_{t,x} v(t,x)\right)^{(2-\gamma)/2} \left(\int_0^L \frac{\partial_x v(t,x)^2}{v(t,x)^{\gamma+2}} dx\right)^{1/2}.$$

The boundedness of v as in (14) is an immediate consequence of the boundedness of (17). Once the boundedness of v is obtained, that of u in H^1 follows just in the same manner as in [5]. We thus conclude (14).

Let (v, u) be a solution of (1), (2), and (3) with initial value (v_0, u_0). If (14) holds, then the orbit of the solution is a precompact set of $C^0 \times C^0$ by the Ascoli-Arzera theorem. In particular, the ω-limit set of the orbit defined by

$$\omega(v_0, u_0) = \bigcap_{n=1}^{\infty} \overline{\{(v(t,\cdot), u(t,\cdot)); t \geq n\}}^{C^0 \times C^0}$$

is nonempty. The following lemma shows that the large time behavior of a bounded solution is under the control of the set of stationary solutions.

Lemma 3 *Assume that* $1 \leq \gamma \leq 2$. *Let* (v, u) *be a bounded solution of (1), (2), and (3) with initial value* (v_0, u_0) *and* $\bar{u} = 0$. *Then, for* $(v_\omega, u_\omega) \in \omega(v_0, u_0)$ *we have* $v_\omega \in C^\infty$, $v_\omega > 0$, $\overline{v_\omega} = \overline{v_0}$, $u_\omega = 0$, *and*

$$\partial_x \left(a v_\omega(x)^{-\gamma}\right) = -\frac{4\pi G}{\overline{v_\omega}} \partial_x \int_0^L K_L(x, y)(v_\omega(y) - \overline{v_\omega}) dy, \tag{19}$$

that is, (v_ω, u_ω) *is a static and stationary solution of (1), (2), and (3) having the average in common with the initial value.*

By **Lemma 2** and the result on boundedness of solutions to the isothermal equations we have $\inf_{t,x} v(t, x) > 0$ and hence $v_\omega > 0$. It is clear that $\overline{v_\omega} = \overline{v_0}$.

We show that $u_\omega = 0$. Choose an increasing sequence t_n, $n = 1, 2, \ldots$, of positive numbers such that $t_n \geq n$ and $\lim_{n\to\infty}(v(t_n,\cdot), u(t_n,\cdot)) = (v_\omega, u_\omega)$ in $C^0 \times C^0$. Since E given by (11) is decreasing, we have

$$\lim_{t\to\infty} E(t) = \lim_{n\to\infty} E(t_n) = \mathscr{E}(v_\omega, u_\omega).$$

We see also that

$$\lim_{n\to\infty} \int_{t_n}^{t_n+1} E(t) dt = \mathscr{E}(v_\omega, u_\omega). \tag{20}$$

Now we consider the limit of

$$\int_{t_n}^{t_n+1} E(t) dt = \int_{t_n}^{t_n+1} \left(\int_0^L \frac{1}{2} u(t, x)^2 dx \right) dt + \int_{t_n}^{t_n+1} \mathscr{E}(v(t,\cdot)) dt$$

term by term. By (13) we see that

$$\lim_{n\to\infty} \int_{t_n}^{t_n+1} \left(\int_0^L \frac{\partial_x u(t, x)^2}{v(t, x)} dx \right) dt = 0. \tag{21}$$

This together with the inequality (18) gives the limit of the first term on the right-hand side:

$$\lim_{n\to\infty}\int_{t_n}^{t_n+1}\left(\int_0^L \frac{1}{2}u(t,x)^2dx\right)dt = 0. \tag{22}$$

In considering the limit of the second term notice that

$$\lim_{n\to\infty}\int_{t_n}^{t_n+1}\left(\int_0^L |v(t,x)-v(t_n,x)|dx\right)dt = 0 \tag{23}$$

holds. This follows from $v(t,x) - v(t_n,x) = \int_{t_n}^t \partial_x u(s,x)ds$, coming from the first equation of (1), and (21). Indeed, for $t \in [t_n, t_n+1]$ we have the following estimate by Schwarz' lemma:

$$\begin{aligned}\int_0^L |v(t,x)-v(t_n,x)|dx &\le \int_{t_n}^{t_n+1}\left(\int_0^L |\partial_x u(s,x)|dx\right)ds \\ &\le \left\{\int_{t_n}^{t_n+1}\left(\int_0^L v(s,x)dx\right)ds\right\}^{1/2}\left\{\int_{t_n}^{t_n+1}\left(\int_0^L \frac{\partial_x u(s,x)^2}{v(s,x)}dx\right)ds\right\}^{1/2} \\ &= \sqrt{L\bar{v}}\left\{\int_{t_n}^{t_n+1}\left(\int_0^L \frac{\partial_x u(s,x)^2}{v(s,x)}dx\right)ds\right\}^{1/2}.\end{aligned}$$

With the use of (23) we can show that

$$\lim_{n\to\infty}\int_{t_n}^{t_n+1}\mathscr{E}(v(t,\cdot))dt = \mathscr{E}(v_\omega).$$

For example, when $\gamma > 1$, from the expression

$$v(t,x)^{1-\gamma}-v(t_n,x)^{1-\gamma} = (1-\gamma)\int_0^1 \{\xi v(t,x)+(1-\xi)v(t_n,x)\}^{-\gamma}\,d\xi\,(v(t,x)-v(t_n,x))$$

we have

$$\begin{aligned}&\int_{t_n}^{t_n+1}\left(\int_0^L \frac{v(t,x)^{1-\gamma}-\bar{v}^{1-\gamma}}{1-\gamma}dx\right)dt - \int_0^L \frac{v(t_n,x)^{1-\gamma}-\bar{v}^{1-\gamma}}{1-\gamma}dx \\ &= \int_0^1\left\{\int_{t_n}^{t_n+1}\left(\int_0^L \{\xi v(t,x)+(1-\xi)v(t_n,x)\}^{-\gamma}(v(t,x)-v(t_n,x))dx\right)dt\right\}d\xi.\end{aligned}$$

Since $\{\xi v(t,x) + (1-\xi)v(t_n,x)\}^{-\gamma} \le (\inf_{t,x} v(t,x))^{-\gamma}$, $0 \le \xi \le 1$, we obtain

$$\lim_{n\to\infty} \int_{t_n}^{t_n+1} \left(\int_0^L \frac{v(t,x)^{1-\gamma} - \bar{v}^{1-\gamma}}{1-\gamma} dx \right) dt = \lim_{n\to\infty} \int_0^L \frac{v(t_n,x)^{1-\gamma} - \bar{v}^{1-\gamma}}{1-\gamma}$$
$$dx = \int_0^L \frac{v_\omega(x)^{1-\gamma} - \bar{v}^{1-\gamma}}{1-\gamma} dx.$$

The limit of the term containing the quadratic form with kernel K_L is handled similarly. Thus,

$$\lim_{n\to\infty} \int_{t_n}^{t_n+1} E(t)dt = \mathscr{E}(v_\omega)$$

holds. Comparing this result with (20), we conclude that $\int_0^L u_\omega(x)^2 dx = 0$, that is, $u_\omega = 0$.

It remains to show that v_ω is smooth and subject to (19). It suffices to verify that the equality (19) holds in the distribution sense. Let t_n, $n = 1, 2, \dots$, be as above. Take a test function ϕ, and a smooth function θ of the real variable with support contained in the interval $(0, 1)$, $\theta \ge 0$, and $\int_0^1 \theta(t)dt = 1$. Multiply the second equation of (1) by $\theta(t-t_n)\phi(x)$ and integrate the both sides of the result over $[t_n, t_n+1] \times [0, L]$. By integration by parts we obtain

$$\begin{aligned}
&-\int_{t_n}^{t_n+1} \theta'(t-t_n) \left(\int_0^L \phi(x)u(t,x)dx \right) dt \\
&-\int_{t_n}^{t_n+1} \theta(t-t_n) \left(\int_0^L \partial_x\phi(x) a v(t,x)^{-\gamma} dx \right) dt \\
&+\mu \int_{t_n}^{t_n+1} \theta(t-t_n) \left(\int_0^L \partial_x\phi(x) \frac{\partial_x u(t,x)}{v(t,x)} dx \right) dt \\
&= \int_{t_n}^{t_n+1} \theta(t-t_n) \left(\int_0^L \partial_x\phi(x) \frac{4\pi G}{\bar{v}} \int_0^L K_L(x,y)(v(t,y)-\bar{v})dydx \right) dt.
\end{aligned}$$

Using (21), (22) and (23) together with $\inf_{t,x} v(t,x) > 0$, we take the limit as $n \to \infty$ term by term. The second term on the left-hand side and the term on the right-hand side are handled in the same manner as shown above. As for the third term on the left-hand side we have the following estimate by Schwarz' lemma:

$$\left| \int_{t_n}^{t_n+1} \theta(t-t_n) \left(\int_0^L \partial_x \phi(x) \frac{\partial_x u(t,x)}{v(t,x)} dx \right) dt \right|$$

$$\leq \left\{ \int_{t_n}^{t_n+1} \theta(t-t_n)^2 \left(\int_0^L \frac{\partial_x \phi(x)^2}{v(t,x)} dx \right) dt \right\}^{1/2} \left\{ \int_{t_n}^{t_n+1} \left(\int_0^L \frac{\partial_x u(t,x)^2}{v(t,x)} dx \right) dt \right\}^{1/2}$$

$$\leq \left(\frac{1}{\inf_{t,x} v(t,x)} \int_0^1 \theta(t)^2 dt \int_0^L \partial_x \phi(x)^2 dx \right)^{1/2} \left\{ \int_{t_n}^{t_n+1} \left(\int_0^L \frac{\partial_x u(t,x)^2}{v(t,x)} dx \right) dt \right\}^{1/2},$$

which shows that the term tends to 0 as $n \to \infty$. Similarly, the first term on the left-hand side tends to 0 as $n \to \infty$. Thus we obtain

$$-\int_0^L \partial_x \phi(x) a v_\omega(x)^{-\gamma} dx = \int_0^L \partial_x \phi(x) \frac{4\pi G}{\overline{v_\omega}} \int_0^L K_L(x,y)(v_\omega(y) - \overline{v_\omega}) dy dx,$$

the equality (19) for v_ω in the distribution sense.

4 Structure and Stability of Stationary Solutions

Consider the stationary problem for (1), (2), and (3):

$$\begin{cases} \partial_x u(x) = 0, \\ \partial_x (a v(x)^{-\gamma}) = -\dfrac{4\pi G}{\bar{v}} \partial_x \displaystyle\int_0^L K_L(x,y)(v(y) - \bar{v}) dy. \end{cases} \tag{24}$$

For a positive parameter V we seek solutions of (24) lying in M_V. Clearly we have $u = 0$. By the change of unknown functions $r(x) = (v(x)/V)^{-\gamma} - 1$, we transform the problem into an equivalent one of finding L-periodic solutions to the following differential equation:

$$\partial_x^2 r(x) + \lambda f(r(x)) = 0, \quad r(x) > -1, \tag{25}$$

with

$$f(r) = 1 - (1+r)^{-1/\gamma}, \quad \lambda = \frac{4\pi G V^\gamma}{a}.$$

It suffices to consider even solutions of (25). To show this let r be an L-periodic solution of (25) and x_0 a critical point of r, i.e., $\partial_x r(x_0) = 0$. Then the functions $r(x + x_0)$ and $r(-x + x_0)$ satisfy (25) with the coincidence of both the values and the derivatives at $x = 0$. It follows from the uniqueness of solutions to the Cauchy problem (25) that $r(x + x_0) = r(-x + x_0)$, i.e., both the functions are even. Thus, any L-periodic solution of (25) is given by an appropriate shift of an even solution.

We seek even L-periodic solutions of (25) in two ways. One relies on the bifurcation theory due to Crandall and Rabinowitz [1], and the other on the first integral of the equation. The latter is enough to study the structure of the whole stationary solutions as shown in **Theorem 1**, but stability analysis of the stationary solutions needs a representation of stationary solutions with a smooth bifurcation curve.

In order to apply the bifurcation theory to (25) we introduce closed subspaces of C^m composed by even functions as $C^m_{\text{even}} = \{r \in C^m; r(-x) = r(x)\}$, $m = 0, 1, 2, \ldots$, and define an infinitely differentiable map H: $\{r \in C^2_{\text{even}}; r > -1\} \times (0, \infty) \to C^0_{\text{even}}$ by

$$H(r, \lambda) = \partial_x^2 r(\cdot) + \lambda f(r(\cdot)).$$

Then the problem is written as a functional equation

$$H(r, \lambda) = 0. \tag{26}$$

It is clear that the constants $(0, \lambda)$, $\lambda > 0$, constitute the so called branch of trivial solutions. Since the Fréchet derivative of H with respect to r is

$$D_r H(r, \lambda)h = \partial_x^2 h + \lambda f'(r(\cdot))h, \quad h \in C^2_{\text{even}},$$

the candidates of the bifurcation points on the branch are obtained by finding such values of λ that the equation $\partial_x^2 h + \lambda f'(0)h = 0$ admits a non-trivial L-periodic solution. From the formula $h(x) = c \cos \sqrt{\lambda f'(0)} x$ with c being a constant, λ must satisfy $\sqrt{\lambda f'(0)} = \frac{2\pi k}{L}$, i.e.,

$$\lambda = \hat{\lambda}^{(k)} \equiv \frac{1}{f'(0)} \left(\frac{2\pi k}{L} \right)^2, \quad k = 1, 2, \ldots.$$

The null space of $D_r H(0, \hat{\lambda}^{(k)})$ is the one dimensional subspace of C^2_{even} spanned by

$$\hat{r}^{(k)}(x) = \cos \frac{2\pi k}{L} x.$$

Furthermore, we have a direct decomposition of C^0_{even}:

$$C^0_{\text{even}} = R(D_r H(0, \hat{\lambda}^{(k)})) \oplus N(D_r H(0, \hat{\lambda}^{(k)})),$$

where $R(D_r H(r, \lambda))$ and $N(D_r H(r, \lambda))$ represent the range and the null space of $D_r H(r, \lambda)$, respectively. This follows from the fact that the equation $\partial_x^2 h +$

$\hat{\lambda}^{(k)} f'(0)h = q$ with $q \in C^0_{\text{even}}$ is solvable if and only if $\int_0^L \hat{r}^{(k)}(x)q(x)dx = 0$ and thus the bounded linear operator on C^0_{even} defined by

$$q \mapsto q - \frac{2}{L}\left(\int_0^L \hat{r}^{(k)}(x)q(x)dx\right)\hat{r}^{(k)}$$

gives a projection onto $R(D_r H(0, \hat{\lambda}^{(k)}))$. Thus the range of $D_r H(0, \hat{\lambda}^{(k)})$ is a closed subspace of C^0_{even} of codimension 1. Finally, it follows from $D_{r\lambda}(r, \lambda)h = f'(r(\cdot))h$, $h \in C^2_{\text{even}}$, that

$$D_{r\lambda}(0, \hat{\lambda}^{(k)})\hat{r}^{(k)} = f'(0)\hat{r}^{(k)} \notin R(D_r H(0, \hat{\lambda}^{(k)})).$$

We are now ready to apply the bifurcation theory of Crandall and Rabinowitz to the equation (26).

Proposition 1 *For a positive integer k there exist a positive constant $\delta^{(k)}$ and a C^∞ curve $s \mapsto (r^{(k)}(\cdot, s), \lambda^{(k)}(s))$ in $C^2_{even} \times (0, \infty)$ defined on $(-\delta^{(k)}, \delta^{(k)})$ satisfying*

$$r^{(k)}(\cdot, 0) = 0, \ \lambda^{(k)}(0) = \hat{\lambda}^{(k)}, \ \partial_s r^{(k)}(\cdot, 0) = \hat{r}^{(k)}, \tag{27}$$

$$\partial_s r^{(k)}(\cdot, s) - \hat{r}^{(k)} \in R(D_r H(0, \hat{\lambda}^{(k)})), \tag{28}$$

and

$$\partial_x^2 r^{(k)}(x, s) + \lambda^{(k)}(s) f(r^{(k)}(x, s)) = 0. \tag{29}$$

Any non-trivial solution (r, λ) of (26) in a suitably small neighborhood of the point$(0, \hat{\lambda}^{(k)})$ lies on the curve. Moreover, we have

$$\partial_s \lambda^{(k)}(0) = 0, \quad \partial_s^2 \lambda^{(k)}(0) = \frac{(1+\gamma)(2-\gamma)}{12\gamma}\left(\frac{2\pi k}{L}\right)^2. \tag{30}$$

The formula (30) shows that the parameter γ determines the direction in which the bifurcation curves extend. In case $1 \leq \gamma < 2$, since $\partial_s \lambda^{(k)}(0) = 0$ and $\partial_s^2 \lambda^{(k)}(0) > 0$, the supercritical bifurcation takes place at every bifurcation point $(0, \hat{\lambda}^{(k)})$, while in case $\gamma > 2$ the subcritical one dose. For the proof of (30) we differentiate the relation (29) repeatedly with respect to the parameter s, and then evaluate the results at $s = 0$. Putting $r_j^{(k)} = \partial_s^j r^{(k)}(\cdot, 0) \in C^2_{\text{even}}$, $\lambda_j^{(k)} = \partial_s^j \lambda^{(k)}(0)$, and using $r_1^{(k)} = \hat{r}^{(k)}$ from (27), we obtain

$$\partial_x^2 r_j^{(k)} + \hat{\lambda}^{(k)} f'(0) r_j^{(k)} = q_j, \tag{31}$$

where

$$
\begin{aligned}
q_1 =&0, \\
q_2 =&-2\lambda_1^{(k)} f'(0)\hat{r}^{(k)} - \hat{\lambda}^{(k)} f''(0) \left(\hat{r}^{(k)}\right)^2, \\
q_3 =&-3\lambda_2^{(k)} f'(0)\hat{r}^{(k)} - 3\lambda_1^{(k)} \left\{f'(0) r_2^{(k)} + f''(0) \left(\hat{r}^{(k)}\right)^2\right\} \\
&- \hat{\lambda}^{(k)} \left\{f'''(0) \left(\hat{r}^{(k)}\right)^3 + 3f''(0)\hat{r}^{(k)} r_2^{(k)}\right\}.
\end{aligned}
$$

Since the equations (31) are indeed solvable, we obtain

$$
\int_0^L \hat{r}^{(k)}(x) q_j(x) dx = 0, \quad j = 2, 3.
$$

From this together with $r_2^{(k)} \in R(D_r H(0, \hat{\lambda}^{(k)}))$ due to (28) we obtain first $\lambda_1^{(k)} = 0$, then $r_2^{(k)}$, and $\lambda_2^{(k)}$ at last.

We proceed to the continuation of the local bifurcation curves. To this end we make use of the relation between the period of a solution and the first integral. The first integral of (25), usually called the energy of the orbit, is given by

$$
\mathscr{I} = \frac{1}{2} \partial_x r(x)^2 + \lambda F(r(x))
$$

with

$$
F(r) = \int_0^r f(s) ds = \begin{cases} r - \log(1+r), & \gamma = 1, \\ r - \dfrac{\gamma}{\gamma - 1}\{(1+r)^{1-1/\gamma} - 1\}, & \gamma > 1. \end{cases}
$$

F is monotone decreasing on $(-1, 0]$ and monotone increasing on $[0, \infty)$, having the limit at either end of the half line:

$$
F(-1+0) = \begin{cases} \infty, & \gamma = 1, \\ \dfrac{1}{\gamma - 1}, & \gamma > 1, \end{cases} \qquad F(\infty) = \infty.
$$

Thus, we can find a unique closed orbit with energy $\mathscr{I}$ if and only if

$$
0 < \mathscr{I} < \begin{cases} \infty, & \gamma = 1, \\ \dfrac{\lambda}{\gamma - 1}, & \gamma > 1. \end{cases}
$$

As is well known, the period l of the orbit with energy $\mathscr{I}$ is given by

$$l = 2\int_{r_{\min}}^{r_{\max}} \frac{dr}{\sqrt{2(\mathscr{I} - \lambda F(r))}},$$

where $r_{\min} < 0$ and $r_{\max} > 0$ are the minimum and the maximum of the solution r, respectively. Note that

$$F(r_{\min}) = F(r_{\max}) = \frac{\mathscr{I}}{\lambda}. \tag{32}$$

By the change of variables $y = \frac{\lambda}{\mathscr{I}}F(r)$ with division of the integral into two, one over $(r_{\min}, 0)$ and the other over $(0, r_{\max})$, we get

$$l = \sqrt{\frac{2}{\lambda}} I_\gamma\left(\sqrt{\frac{\mathscr{I}}{\lambda}}\right) \tag{33}$$

with I_γ given by (8). By differentiation and limit under the integral sign, we obtain

Lemma 4 *Assume* $1 \le \gamma < 2$. *Then,* $I'_\gamma(\theta) > 0$. *Moreover we have*

$$I_\gamma(+0) = \sqrt{2\gamma}\pi, \qquad \begin{cases} I_\gamma(\infty) = \infty, & \gamma = 1, \\ I_\gamma((\gamma-1)^{-1/2} - 0) < \infty, & 1 < \gamma < 2. \end{cases}$$

This shows the monotonic relation between the period and the energy with lower and upper limits. Thus, a necessary and sufficient condition for the existence and uniqueness of l-periodic orbits is established:

$$\sqrt{\frac{4\pi^2\gamma}{\lambda}} < l < \begin{cases} \infty, & \gamma = 1, \\ \sqrt{\frac{2}{\lambda}} I_\gamma((\gamma-1)^{-1/2} - 0), & 1 < \gamma < 2. \end{cases} \tag{34}$$

Recalling $\lambda = \frac{4\pi G V^\gamma}{a}$, we obtain the assertion of **Theorem 1** immediately.

Let us comment on the relation between the above two results concerning the structure of the set of solutions to (26). For a positive integer k let λ be a positive parameter satisfying (34) with $l = L/k$, that is,

$$\gamma\left(\frac{2\pi k}{L}\right)^2 < \lambda < \begin{cases} \infty, & \gamma = 1, \\ 2\left(\frac{I_\gamma((\gamma-1)^{-1/2} - 0)k}{L}\right)^2, & 1 < \gamma < 2. \end{cases} \tag{35}$$

For such a λ the equation (25) admits a unique orbit with period L/k. Let r be an even solution having the orbit. The whole of such pairs (r, λ) together with the trivial solution $(0, \gamma\left(\frac{2\pi k}{L}\right)^2)$ constitutes a global solution curve of (26) that continues the local bifurcation curve $(r^{(k)}, \lambda^{(k)})$ of **Proposition 1** in the following sense. From (32) and (33) together with **Lemma 4** we have

$$F(r_{\min}) = F(r_{\max}) = \left(I_\gamma{}^{-1}\left(\frac{L}{k}\sqrt{\frac{\lambda}{2}}\right)\right)^2.$$

This implies that both $r_{\min}$ and $r_{\max}$ depend continuously on the parameter λ. In particular, as $\lambda \to \gamma\left(\frac{2\pi k}{L}\right)^2 + 0$, we have $r_{\min}, r_{\max} \to 0$ because $F(r_{\min}) = F(r_{\max}) \to \left(I_\gamma{}^{-1}(\sqrt{2\gamma}\pi + 0)\right)^2 = 0$. Therefore the global solution curve lies on the bifurcation curve $(r^{(k)}, \lambda^{(k)})$ in a small neighborhood of the bifurcation point $(0, \hat{\lambda}^{(k)})$. We see also that an orbit with period other than L/k, if exists near $\lambda = \gamma\left(\frac{2\pi k}{L}\right)^2$, keeps a definite amplitude there. Thus, in a small neighborhood of the bifurcation point $(0, \hat{\lambda}^{(k)})$ the bifurcation curve $(r^{(k)}, \lambda^{(k)})$ in turn must lie on the global solution curve and hence the two curves overlap with each other.

We next consider the behavior of solutions to (1), (2), and (3) near a stationary solution. Specifically, we are interested in the stability of the stationary solution. Let $(\tilde{v}, 0)$ be a stationary solution of (1), (2), and (3) with $\bar{\tilde{v}} = V$, and (v, u) a solution of (1), (2), and (3) with $\bar{v} = V$ under consideration. We introduce the disturbance from the stationary solution as

$$\phi(t, x) = v(t, x) - \tilde{v}(x), \quad \psi(t, x) = u(t, x). \tag{36}$$

Supposing that the disturbance is small enough in amplitude, we expand the energy form (5), and (6) with respect to the disturbance. Using the equation (24) for $(\tilde{v}, 0)$, we obtain

$$\begin{aligned} E(t) &= \mathscr{E}(v(t, \cdot), u(t, \cdot)) \\ &= \mathscr{E}(\tilde{v} + \phi(t, \cdot), \psi(t, \cdot)) \\ &= \mathscr{E}(\tilde{v}) + \frac{1}{2}\|\psi(t, \cdot)\|_{L^2}^2 + \frac{1}{2}Q[\phi(t, \cdot)] + \mathscr{O}(\|\phi(t, \cdot)\|_{L^\infty})\|\phi(t, \cdot)\|_{L^2}^2, \end{aligned} \tag{37}$$

where $\|\cdot\|_{L^\infty}$ is the supremum norm, and Q is the quadratic form on the Hilbert space $\mathscr{H} = \{\varphi \in L^2; \bar{\varphi} = 0\}$ defined by

$$Q[\varphi] = \int_0^L a\frac{\gamma\varphi(x)^2}{\tilde{v}(x)^{\gamma+1}}dx - \frac{4\pi G}{V}\int_0^L\int_0^L K_L(x, y)\varphi(x)\varphi(y)dxdy. \tag{38}$$

The expression (37) together with the decrease of E suggests that the stability of the stationary solution is closely related to the "sign" of the quadratic form. Thus we are led to the notion of linearized stability.

Definition 1

(i) Suppose there exists a positive constant δ such that

$$Q[\varphi] \geq \delta \|\varphi\|_{L^2}^2, \quad \varphi \in \mathscr{H}.$$

Then, we can expect possible decay in L^2 of any small disturbance. In this case we call the stationary solution $(\tilde{v}, 0)$ to be linearly stable.

(ii) Suppose the quadratic form Q admits a negative value, i.e., $Q[\varphi_0] < 0$ for some $\varphi_0 \in \mathscr{H}$. Without loss of generality we may assume that φ_0 is smooth. Since

$$\mathscr{E}(\tilde{v} + \varepsilon\varphi_0, 0) = \mathscr{E}(\tilde{v}) + \frac{1}{2}\varepsilon^2 Q[\varphi_0] + \mathscr{O}(|\varepsilon|^3) \quad \text{for small } |\varepsilon|,$$

and hence the energy form takes a value less than its value at the stationary solution in any small neighborhood of that stationary solution, there do exist some arbitrarily small disturbances with no decay but possible growth. In this case we call the stationary solution $(\tilde{v}, 0)$ to be linearly unstable.

In examining the linearized stability it is of great use that the quadratic form Q is expressed as

$$Q[\varphi] = \frac{a}{V^{\gamma+1}}(T\varphi, \varphi)_{L^2},$$

where T is the self-adjoint operator on $\mathscr{H}$ given by

$$(T\varphi)(x) = \frac{\gamma\varphi(x)}{(1+\tilde{w}(x))^{\gamma+1}} - \frac{1}{L}\int_0^L \frac{\gamma\varphi(x)}{(1+\tilde{w}(x))^{\gamma+1}}dx - \lambda \int_0^L K_L(x,y)\varphi(y)dy \tag{39}$$

with $\tilde{w} = \tilde{v}/V - 1$ and $\lambda = \frac{4\pi G V^\gamma}{a}$. From the fact that $\inf_{\|\varphi\|_{L^2}=1}(T\varphi, \varphi)_{L^2}$ gives the lower bound of $\sigma(T)$, the spectrum of T, the study on the sign of Q is reduced to the spectral analysis of T. We remark that $\inf \sigma(T) \leq \gamma(1 + \max \tilde{w})^{-\gamma-1}$ since we have $(T\varphi, \varphi)_{L^2} \leq \int_0^L \frac{\gamma\varphi(x)^2}{(1+\tilde{w}(x))^{\gamma+1}}dx$ for $\varphi \in \mathscr{H}$. Furthermore, notice that in the region below $\gamma(1+\max \tilde{w})^{-\gamma-1}$ the spectrum consists of eigenvalues of T. To show this let us consider an equation in $\mathscr{H}$ with parameter $\Lambda < \gamma(1 + \max \tilde{w})^{-\gamma-1}$:

$$T\varphi - \Lambda\varphi = \psi,$$

or equivalently, $P_\Lambda\varphi - \lambda K_L\varphi = \psi$ with

$$(P_\Lambda\varphi)(x) = \left\{\frac{\gamma}{(1+\tilde{w}(x))^{\gamma+1}} - \Lambda\right\}\varphi(x) - \frac{1}{L}\int_0^L \frac{\gamma\varphi(x)}{(1+\tilde{w}(x))^{\gamma+1}}dx,$$

$$(K_L\varphi)(x) = \int_0^L K_L(x,y)\varphi(y)dy.$$

Noting that P_Λ is a positive definite self-adjoint operator on $\mathscr{H}$, we transform the equation into

$$(I - \lambda {P_\Lambda}^{-1}K_L)\varphi = {P_\Lambda}^{-1}\psi. \tag{40}$$

As is well known, K_L is a compact operator on $\mathscr{H}$, and so is $\lambda {P_\Lambda}^{-1}K_L$. By the Riesz-Schauder theory either the equation (40) has a unique solution φ for any $\psi \in \mathscr{H}$ or there exists a non-trivial element φ of $\mathscr{H}$ satisfying $(I - \lambda {P_\Lambda}^{-1}K_L)\varphi = 0$. If $\Lambda \in \sigma(T)$, then the latter must hold, showing that Λ is an eigenvalue of T with eigenvector φ.

We now proceed to the eigenvalue problem for T. For a trivial solution the problem $T\varphi = \Lambda\varphi$ is equivalent to finding L-periodic solutions of the differential equation $\gamma\partial_x^2\varphi + \lambda\varphi = \Lambda\partial_x^2\varphi$. According to the formula $\varphi(x) = c_1\cos\sqrt{\frac{\lambda}{\gamma-\Lambda}}x + c_2\sin\sqrt{\frac{\lambda}{\gamma-\Lambda}}x$ with c_1, c_2 being constants the equation admits a non-trivial solution with period L if and only if $\sqrt{\frac{\lambda}{\gamma-\Lambda}} = \frac{2\pi k}{L}$, i.e., $\Lambda = \gamma - \frac{\lambda L^2}{4\pi^2k^2}$, $k = 1, 2, \ldots$. Therefore the lower bound of T is $\gamma - \frac{\lambda L^2}{4\pi^2}$.

To study the eigenvalue problem for T corresponding to a non-trivial stationary solution some preliminary observations are in order. One is concerned with the null space of T. Let $(\tilde{v}, 0)$ be a non-trivial solution of (24) with $\bar{\tilde{v}} = V$. Then $\tilde{w} = \tilde{v}/V - 1$ satisfies the following equivalent to the second equation of (24):

$$\Theta(\tilde{w},\lambda) = 0, \quad \lambda = \frac{4\pi G V^\gamma}{a}, \tag{41}$$

where $\Theta(w,\lambda)$ is the mapping from $\{w \in C^0; \bar{w} = 0, w > -1\} \times (0,\infty)$ into $\{w \in C^0; \bar{w} = 0\}$ given by

$$\Theta(w,\lambda)(x) = -\frac{1}{(1+w(x))^\gamma} + \frac{1}{L}\int_0^L \frac{dx}{(1+w(x))^\gamma} - \lambda\int_0^L K_L(x,y)w(y)dy.$$

We know that for $\alpha \in \mathbf{R}$, $(\tilde{v}(\cdot-\alpha), 0)$ solves (24) and thus $\Theta(\tilde{w}(\cdot-\alpha),\lambda) = 0$ holds. Differentiating this relation with respect to α and evaluating the result at $\alpha = 0$, we obtain

$$\frac{\gamma\tilde{w}'(x)}{(1+\tilde{w}(x))^{\gamma+1}} - \frac{1}{L}\int_0^L \frac{\gamma\tilde{w}'(x)}{(1+\tilde{w}(x))^{\gamma+1}}dx - \lambda\int_0^L K_L(x,y)\tilde{w}'(y)dy = 0,$$

that is, $T\tilde{w}' = 0$. This shows that T has a non-trivial null space with an eigenvector $\tilde{w}' \neq 0$, and hence the lower bound of T is a nonpositive eigenvalue.

Another observation is the invariance of the point spectrum under the shift of the stationary solution. Let Λ be an eigenvalue of T with eigenvector φ. From the equation $T\varphi = \Lambda\varphi$ we have

$$\frac{\gamma\varphi(x-\alpha)}{(1+\tilde{w}(x-\alpha))^{\gamma+1}} - \frac{1}{L}\int_{\alpha}^{L+\alpha} \frac{\gamma\varphi(x-\alpha)}{(1+\tilde{w}(x-\alpha))^{\gamma+1}}dx - \lambda$$
$$\int_{\alpha}^{L+\alpha} K_L(x-\alpha, y-\alpha)\varphi(y-\alpha)dy = \Lambda\varphi(x-\alpha)$$

for any $\alpha \in \mathbf{R}$. Noting that $K_L(x, y)$ is a function of $x - y$, by the L-periodicity of $\tilde{w}$, φ and $K_L(x, \cdot)$ we obtain $T^{\alpha}\varphi(\cdot - \alpha) = \Lambda\varphi(\cdot - \alpha)$ with T^{α} corresponding to the stationary solution $(\tilde{v}(\cdot - \alpha), 0)$ as in (39). This shows that Λ is an eigenvalue of T^{α} with eigenvector $\varphi(\cdot - \alpha)$. Thus the point spectrum of T^{α} coincides with that of T for any $\alpha \in \mathbf{R}$ with the correspondence of associated eigenspaces specified as above. This allows us to study the lower bound of T after a favorable shift of $\tilde{v}$.

With the above considerations in mind, let us set about the observation of every branch of the stationary solutions. For a positive integer k let λ be a parameter satisfying (35). Let r_λ denote a unique solution of (25) with L/k-periodic orbit, taking its minimum at $x = 0$. The relation $\tilde{w}_\lambda = (1 + r_\lambda)^{-1/\gamma} - 1$ gives the correspondence between r_λ and a unique solution $\tilde{w}_\lambda$ of (41) having the least period L/k and the maximum at $x = 0$. The continuous dependence on λ of both the minimum and the maximum of r_λ shown above means that r_λ depends continuously on λ in the space C^0_{even} up to the left endpoint of the interval, and so dose $\tilde{w}_\lambda$. The operator (39) with $\tilde{w} = \tilde{w}_\lambda$ is denoted by T_λ. Then the lower bound of T_λ as well as $\tilde{w}_\lambda$ depends continuously on λ. Recall that T_λ has the eigenvalue 0 with an eigenvector $\tilde{w}'_\lambda$, being an odd function. The eigenvalue is in fact simple. To show this we notice that an element φ of the null space of T_λ is subject to the following second order linear differential equation:

$$\partial_x^2\{\gamma(1+\tilde{w}_\lambda(x))^{-\gamma-1}\psi(x)\} - \lambda\psi(x) = 0.$$

We express φ as a linear combination of the odd solution $\tilde{w}'_\lambda$ and the even solution ψ with $\psi(0) = 1$, which solves the variational equation of the Cauchy problem

$$\begin{cases} \partial_x^2\{-(1+w(x))^{-\gamma}\} - \lambda w(x) = 0, \\ w(0) = \xi, \quad \partial_x w(0) = 0, \end{cases}$$

taken at $\xi = \tilde{w}_\lambda(0)$. Using the result $I'_\gamma(\theta) > 0$ of **Lemma 4**, we can prove that the variational equation by no means admits a non-trivial solution with period L. Thus φ is proportional to $\tilde{w}'_\lambda$.

Thanks to the simplicity of the eigenvalue 0 and the continuity of the lower bound of T_λ we have an important information in hand on the sign of the lower bound. Once the lower bound takes a negative value somewhere on the interval (35) of λ, it remains negative over the interval, otherwise the negative eigenvalue giving the lower bound would meet the eigenvalue 0 at some point on the interval, contradicting the simplicity of the eigenvalue 0. Here we notice that in the limit $\lambda \to \gamma\left(\frac{2\pi k}{L}\right)^2 + 0$, the lower bound of T_λ approaches that of the operator corresponding to the trivial solution $\left(\left(\frac{a\gamma\pi k^2}{GL^2}\right)^{1/\gamma}, 0\right)$, the operator with lower bound $\gamma(1-k^2)$. This implies that the operators corresponding to the stationary solutions found on the bifurcation branches except for the first one have negative lower bounds.

In order to determine the sign of the lower bound of T_λ in case $k = 1$, we make use of a result of Crandall and Rabinowitz [2] on the perturbation of a simple eigenvalue of a linearized operator along a smooth bifurcation curve. See also [4], Proposition I. 7. 2. Let $(r^{(1)}(s), \lambda^{(1)}(s))$ be the solution of (26) as in **Proposition 1**. Put $w(s) = (1 + r^{(1)}(s))^{-1/\gamma} - 1$, $\lambda(s) = \lambda^{(1)}(s)$. Then the family $(w(s), \lambda(s))$ gives a smooth solution curve of $\Theta(w, \lambda) = 0$. Recall that for $|s| \neq 0$ sufficiently small $\lambda(s) > \gamma\left(\frac{2\pi}{L}\right)^2$ holds, and hence $w(s)$ is given by an appropriate shift of $\tilde{w}_{\lambda(s)}$. The extension of the Fréchet derivative $D_w\Theta(w(s), \lambda(s))$ onto $\mathscr{H}$ coincides with $T^\alpha_{\lambda(s)}$ for some α. Thus, an eigenvalue of $D_w\Theta(w(s), \lambda(s))$ is also an eigenvalue of $T_{\lambda(s)}$. In view of this observation we consider the behavior of the perturbed eigenvalue of $D_w\Theta(w(s), \lambda(s))$ due to Crandall and Rabinowitz, parametrized by s and issuing from 0 at $s = 0$. Calculating the direction in which the eigenvalue moves with the aid of (30), we see that there exists a positive eigenvalue of T_λ for λ sufficiently close to $\gamma\left(\frac{2\pi}{L}\right)^2$, tending to 0 as $\lambda \to \gamma\left(\frac{2\pi}{L}\right)^2 + 0$. This assures that the lower bound of T_λ is the eigenvalue 0, otherwise the hypothetical negative eigenvalue of T_λ, as well as the eigenvalue 0 and the positive one, would meet 0 at $\lambda = \gamma\left(\frac{2\pi}{L}\right)^2$, which contradicts the fact that the null space of the operator corresponding to the trivial solution $\left(\left(\frac{a\gamma\pi}{GL^2}\right)^{1/\gamma}, 0\right)$ is two dimensional.

A similar argument shows that zero is an isolated eigenvalue of T_λ. Summing up, we obtain the following result.

Proposition 2 *Assume* $1 \leq \gamma < 2$. *Let* $(\tilde{v}^{(j)}, 0)$ *be the stationary solution of (1), (2), and (3) as in* ***Theorem 1***.

(i) *For* $\tilde{v} = V$, *we have* $\inf \sigma(T) = \gamma - \frac{GL^2 V^\gamma}{a\pi}$.
(ii) *For* $\tilde{v} = \tilde{v}^{(1)}$, *we have* $\inf \sigma(T) = 0$ *and* $\inf(\sigma(T) \setminus \{0\}) > 0$, *while for* $\tilde{v} = \tilde{v}^{(j)}$, $j = 2, 3, \ldots,$ $\inf \sigma(T) < 0$.

So far we leave out the case in which the quadratic form (38) corresponding to a stationary solution is nonnegative and degenerate. As **Proposition 2** shows, this is true of the stationary solutions found on the first bifurcation branch. Recall that the whole of these stationary solutions constitutes a smooth curve on M_V, and that the degeneracy of the quadratic forms takes place only in the directions tangent to the curve at the respective stationary solutions. With the aid of this geometrical and spectral structure we show that the set of those stationary solutions enjoys a revision of the linear stability.

Write $\tilde{v}^{(1)}$ as $\tilde{v}$ for simplicity, and put

$$S_V^{(1)} = \{(\tilde{v}(\cdot - \alpha), 0); 0 \le \alpha < L\}.$$

Instead of considering a disturbance as in (36) with a stationary solution fixed on $S_V^{(1)}$, we introduce one from the set $S_V^{(1)}$ as follows:

$$\phi(t,x) = v(t,x) - \tilde{v}(x - \alpha_{\min}), \quad \psi(t,x) = u(t,x),$$

where we take a shift $\alpha_{\min} \in [0, L)$ so that

$$\|v(t,\cdot) - \tilde{v}(\cdot - \alpha_{\min})\|_{L^2} = \min_{\alpha} \|v(t,\cdot) - \tilde{v}(\cdot - \alpha)\|_{L^2}.$$

Such a shift is uniquely determined as long as the state $(v(t,\cdot), u(t,\cdot))$ stays in a small neighborhood of $S_V^{(1)}$. Noting that the value $\mathscr{E}(\tilde{v}(\cdot - \alpha))$ is independent of α, we expand the energy form around the stationary solution $(\tilde{v}(\cdot - \alpha_{\min}), 0)$ to obtain

$$E(t) = \mathscr{E}(\tilde{v}) + \frac{1}{2}\|\psi(t,\cdot)\|_{L^2}^2 + \frac{1}{2}Q^{\alpha_{\min}}[\phi(t,\cdot)] + \mathscr{O}(\|\phi(t,\cdot)\|_{L^\infty})\|\phi(t,\cdot)\|_{L^2}^2, \tag{42}$$

where $Q^{\alpha_{\min}}$ is the quadratic form corresponding to the stationary solution $(\tilde{v}(\cdot - \alpha_{\min}), 0)$. Now evaluate at $\alpha = \alpha_{\min}$ the both sides of the formula

$$\frac{d}{d\alpha}\|v(t,\cdot) - \tilde{v}(\cdot - \alpha)\|_{L^2}^2 = 2\int_0^L (v(t,x) - \tilde{v}(x-\alpha))\tilde{v}'(x-\alpha)dx.$$

Then, from $\tilde{v} = V(1 + \tilde{w})$ we have $(\phi(t,\cdot), \tilde{w}'(\cdot - \alpha_{\min}))_{L^2} = 0$, which shows that $\phi(t,\cdot)$ is orthogonal to the null space of the operator $T^{\alpha_{\min}}$ associated with the quadratic form $Q^{\alpha_{\min}}$. By **Proposition 2** we notice that $\kappa = \inf(\sigma(T^\alpha) \setminus \{0\})$ is a positive number independent of α. Since

$$Q^{\alpha_{\min}}[\phi(t,\cdot)] = \frac{a}{V^{\gamma+1}}(T^{\alpha_{\min}}\phi(t,\cdot), \phi(t,\cdot))_{L^2} \ge \frac{a\kappa}{V^{\gamma+1}}\|\phi(t,\cdot)\|_{L^2}^2,$$

the expression (42) suggests possible decay in L^2 of any small disturbance as a result of the decrease of the energy form. In this sense we call the set $S_V^{(1)}$ of stationary solutions to be linearly stable.

5 Existence of Unbounded Solutions

In Sect. 3 we obtained two alternatives on the asymptotics of a solution to (1), (2), and (3); the solution approaches the set of stationary solutions, or it is unbounded, i.e., $\sup_{t,x} v(t,x) = \infty$. Our task is finding out some initial conditions of the solution that rule out the former. An obvious, probably not unique, approach is the use of the decreasing property of the energy form. Suppose there exists a state on M_V at which the energy form takes a value smaller than those of the energy form evaluated at the stationary solutions on M_V. Let us consider the orbit passing such a state. Then the decrease of the energy form along the orbit assures that the orbit is apart from the set of the stationary solutions. Thus the first alternative is clearly rejected. This observation gives us the idea of providing an initial condition for unbounded solutions in terms of the energy form. Needless to say, we should be careful to make the condition meaningful. Based on this idea, we first examine at which stationary solution the energy form takes the minimal value, and then look for a desired state around the stationary solution minimizing the values of the energy form.

Assume $1 \leq \gamma < 2$. We further assume that the stationary problem for (1), (2), and (3) has a non-trivial solution on M_V. For $j = k_{\min}, \ldots, k_{\max}$ choose a stationary solution $(\tilde{v}^{(j)}, 0) \in M_V$ with least period L/j as in **Theorem 1**, and put

$$S_V = \{(\tilde{v}^{(j)}, 0); j = k_{\min}, \ldots, k_{\max}\} \cup \{(V, 0)\}.$$

To read where the energy form attains its minimum on S_V it suffices to compare the values $\mathscr{E}(\tilde{v}^{(j)})$, $j = k_{\min}, \ldots, k_{\max}$, and $\mathscr{E}(V) = 0$ with each other. In this procedure it is of great use to consider how the values except for $\mathscr{E}(V)$ depend on the least periods of the corresponding stationary solutions. To this end we introduce the following function with respect to the periods of stationary solutions:

$$\varepsilon(l) = \int_0^l \int_V^{\tilde{v}^l(x)} (p(V) - p(\xi)) d\xi dx - \frac{2\pi G}{V} \int_0^l \int_0^l K_l(x,y)(\tilde{v}^l(x) - V)(\tilde{v}^l(y) - V) dx dy, \tag{43}$$

where $p(v) = av^{-\gamma}$ as before, and $(\tilde{v}^l, 0)$ is the non-trivial solution of the stationary problem (24) parametrized by $L = l$ with $\tilde{v}^l$ having the average V, the least period l, and the maximum at $x = 0$. As established in (34), $\tilde{v}^l$ as well as $\varepsilon(l)$ is well defined for l with

$$\left(\frac{a\gamma\pi}{GV^\gamma}\right)^{1/2} < l < \begin{cases} \infty, & \gamma = 1, \\ \dfrac{I_\gamma((\gamma-1)^{-1/2} - 0)}{\sqrt{2\gamma\pi}} \left(\dfrac{a\gamma\pi}{GV^\gamma}\right)^{1/2}, & 1 < \gamma < 2. \end{cases}$$

In terms of this function $\mathscr{E}(\tilde{v}^{(j)})$ is expressed as follows.

Recall (6). In case $\gamma = 1$ we have

$$\mathscr{E}(v) = \int_0^L a\left(\frac{v(x)-\bar{v}}{\bar{v}} - \log\frac{v(x)}{\bar{v}}\right)dx$$
$$-\frac{2\pi G}{\bar{v}}\int_0^L\int_0^L K_L(x,y)(v(x)-\bar{v})(v(y)-\bar{v})dxdy. \tag{44}$$

We consider only this case since the case $1 < \gamma < 2$ is handled quite similarly with the use of (10) instead of (44). Put $l_j = L/j$. Notice that $\mathscr{E}(\tilde{v}^{(j)}) = \mathscr{E}(\tilde{v}^{l_j})$. In the expression of $\mathscr{E}(\tilde{v}^{l_j})$ we divide every integral on the interval $[0, L]$ into the integrals on the subintervals $[ml_j, (m+1)l_j]$, $m = 0, \ldots, j-1$, and rewrite every piece as an integral on $[0, l_j]$ by change of variables. By periodicity of $\tilde{v}^{l_j}$ we have

$$\mathscr{E}(\tilde{v}^{(j)}) = j\int_0^{l_j} a\left(\frac{\tilde{v}^{l_j}(x)-V}{V} - \log\frac{\tilde{v}^{l_j}(x)}{V}\right)dx$$
$$-\frac{2\pi G}{V}\int_0^{l_j}\int_0^{l_j}\sum_{m,n=0}^{j-1} K_L(x+ml_j, y+nl_j)(\tilde{v}^{l_j}(x)-V)(\tilde{v}^{l_j}(y)-V)dxdy.$$

From the formula (4) of $K_L(x,y)$ we see by arithmetics that

$$\sum_{m,n=0}^{j-1} K_L(x+ml_j, y+nl_j) = jK_{l_j}(x,y), \quad 0 \le x, y \le l_j.$$

Thus we obtain the following:

Lemma 5 *For $j = k_{\min}, \ldots, k_{\max}$ we have*

$$\mathscr{E}(\tilde{v}^{(j)}) = j\varepsilon(L/j). \tag{45}$$

Rewriting (45) as

$$\mathscr{E}(\tilde{v}^{(j)}) = L\frac{\varepsilon(L/j)}{L/j},$$

we are forced to study the behavior of the function $\varepsilon(l)/l$. Put $\tilde{v}_0^l = \tilde{v}^l(0)$. Recalling the relation between the two equations (24) and (25), we have

$$\tilde{v}_0^l = V(1 + r_{\min}^l)^{-1/\gamma}, \tag{46}$$

where $r^l_{\min} < 0$ is given by the relation

$$F(r^l_{\min}) = \left(I_\gamma^{-1} \left(l \sqrt{\frac{\lambda}{2}} \right) \right)^2, \quad \lambda = \frac{4\pi G V^\gamma}{a}. \tag{47}$$

See (32) and (33). This implies that $\tilde{v}^l_0$ depends smoothly on the parameter l, and so do $\tilde{v}^l$ and $\varepsilon(l)$. Taking the derivative of $\varepsilon(l)$, we rewrite the result using $\tilde{v}^l(l) = \tilde{v}^l_0$, $\int_0^l \partial_l \tilde{v}^l(x)dx = V - \tilde{v}^l_0$ from $\int_0^l \tilde{v}^l(x)dx = Vl$, and the symmetry of the Green kernel $K_l(x, y)$. Then we have

$$\begin{aligned}
\varepsilon'(l) = &- \int_V^{\tilde{v}^l_0} a\xi^{-\gamma} d\xi \\
&- \int_0^l a \frac{\partial_l \tilde{v}^l(x)}{\tilde{v}^l(x)^\gamma} dx - \frac{4\pi G}{V} \int_0^l \int_0^l K_l(x, y) \partial_l \tilde{v}^l(x)(\tilde{v}^l(y) - V) dxdy \\
&- \frac{4\pi G}{V} \int_0^l K_l(l, y)(\tilde{v}^l(y) - V) dy (\tilde{v}^l_0 - V) \\
&- \frac{2\pi G}{V} \int_0^l \int_0^l \partial_l K_l(x, y)(\tilde{v}^l(x) - V)(\tilde{v}^l(y) - V) dxdy.
\end{aligned}$$

The sum of the second and the third terms on the right-hand side is

$$-\frac{1}{l} \int_0^l \frac{a}{\tilde{v}^l(x)^\gamma} dx \int_0^l \partial_l \tilde{v}^l(x) dx = \frac{1}{l} \int_0^l \frac{a}{\tilde{v}^l(x)^\gamma} dx \, (\tilde{v}^l_0 - V),$$

which turns out to be $a(\tilde{v}^l_0 - V)/(\tilde{v}^l_0)^\gamma$ with addition of the forth term. This comes from the following equation equivalent to the second one of (24) with $L = l$:

$$-\frac{a}{\tilde{v}^l(x)^\gamma} + \frac{1}{l} \int_0^l \frac{a}{\tilde{v}^l(x)^\gamma} dx - \frac{4\pi G}{V} \int_0^l K_l(x, y)(\tilde{v}^l(y) - V) dy = 0.$$

Since $\tilde{v}^l$ is axially symmetric with respect to $x = l/2$, the last term on the right-hand side vanishes in view of

$$\begin{aligned}
\partial_l K_l(x, y) &= -\frac{(x - y)^2}{2l^2} + \frac{1}{12} = -\frac{1}{2l^2} \left\{ \left(x - \frac{l}{2} \right)^2 + \left(y - \frac{l}{2} \right)^2 \right. \\
&\quad \left. -2 \left(x - \frac{l}{2} \right) \left(y - \frac{l}{2} \right) \right\} + \frac{1}{12}, \quad 0 \leq x, y \leq l.
\end{aligned}$$

Thus we obtain

$$\varepsilon'(l) = a\left\{-\int_V^{\tilde{v}_0^l} \xi^{-\gamma}\,d\xi + \frac{\tilde{v}_0^l - V}{(\tilde{v}_0^l)^\gamma}\right\},$$

and hence

$$\varepsilon''(l) = -a\gamma\frac{\partial_l\tilde{v}_0^l\,(\tilde{v}_0^l - V)}{(\tilde{v}_0^l)^{\gamma+1}}.$$

Applying the result of **Lemma 4** to (46) with (47), we see that $\partial_l\tilde{v}_0^l > 0$. This together with $\tilde{v}_0^l > V$ gives $\varepsilon''(l) < 0$. Again by **Lemma 4** we have $\tilde{v}_0^l \to V$ as $l \to \left(\frac{a\gamma\pi}{GV^\gamma}\right)^{1/2} + 0$. Therefore both $\varepsilon(l)$ and $\varepsilon'(l)$ tend to 0 as $l \to \left(\frac{a\gamma\pi}{GV^\gamma}\right)^{1/2} + 0$. Combining the results together with $(\varepsilon(l)/l)' = (l\varepsilon'(l) - \varepsilon(l))/l^2$ and $(l\varepsilon'(l) - \varepsilon(l))' = l\varepsilon''(l) < 0$, we obtain

Lemma 6 *We have* $(\varepsilon(l)/l)' < 0$ *and* $\varepsilon(l) < 0$.

The following is an immediate consequence of **Lemma 6**.

Proposition 3 *For* $j, j' = k_{\min}, \ldots, k_{\max}$ *with* $j < j'$, *we have*

$$\mathscr{E}(\tilde{v}^{(j)}) < \mathscr{E}(\tilde{v}^{(j')}) < \mathscr{E}(V) = 0. \tag{48}$$

In particular, $\mathscr{E}(\tilde{v}^{(k_{\min})})$ *is minimal amongst the values of the energy form on* S_V.

We are now in position to seek on M_V a state lower than any of the stationary solutions in comparison of values of the energy form. It depends on the stability of a stationary solution whether the energy form admits, in a neighborhood of the stationary solution, a value smaller than that of the energy form evaluated at the stationary solution. If the stationary solution or the set of stationary solutions is linearly stable in the sense given in Sect. 4, then the value of the energy form evaluated at the stationary solution is in fact a local minimum. We therefore miss a state with desired value of the energy form if we choose a neighborhood of the stationary solution sufficiently small. To the contrary, the linear instability of the stationary solution assures the existence of such a state in any small neighborhood of the stationary solution.

Combining all the auxiliary considerations together, we enter into the details of the statement of the main theorem. Assume $1 < \gamma < 2$. When $V < \left(\frac{a\gamma\pi}{GL^2}\right)^{1/\gamma}$, the trivial solution $(V, 0)$ is the unique stationary solution on M_V, which is linearly stable. Similarly, when $\left(\frac{a\gamma\pi}{GL^2}\right)^{1/\gamma} < V < \left(\frac{al_\gamma((\gamma-1)^{-1/2}-0)^2}{2\pi GL^2}\right)^{1/\gamma}$, the set of stationary solutions $(\tilde{v}^{(1)}(\cdot - \alpha), 0)$, $0 \le \alpha < L$, found on the first bifurcation branch gives the minimal value of the energy form on S_V, being linearly stable as well. In both cases, it is true that the stationary solutions give local minimums of the energy form on M_V, but whether they give global minimums on the whole M_V in the respective cases of V is beyond the scope of the linear stability analysis. On this point the situation is very complicated as compared with that of the isothermal case $\gamma = 1$. In this case, because of the boundedness of solutions to (1), (2), and (3) the energy form must attain its global minimum on M_V either at the trivial solution or at the set of stationary solutions found on the first bifurcation branch. In the isentoropic case likewise, it is probably hard to find somewhere on M_V states lower than the stationary solutions in comparison of values of the energy form, however, at the present we are obliged to leave the case $V < \left(\frac{aI_\gamma((\gamma-1)^{-1/2}-0)^2}{2\pi GL^2}\right)^{1/\gamma}$, including the even more subtle case $V = \left(\frac{a\gamma\pi}{GL^2}\right)^{1/\gamma}$, for further studies. On the other hand the case $V \geq \left(\frac{aI_\gamma((\gamma-1)^{-1/2}-0)^2}{2\pi GL^2}\right)^{1/\gamma}$ is rather easy to handle. For such a V the stationary solutions on M_V are all linearly unstable. In particular, there dose exist on M_V a state with value of the energy form less than the minimal one, $\mathscr{E}(\tilde{v}^{(k_{\min})})$ if it is meaningful or else $\mathscr{E}(V) = 0$. Now we arrive at the conclusion of **Theorem 2**.

References

1. Crandall, M.G., Rabinowitz, P.H.: Bifurcation from simple eigenvalues. J. Funct. Anal. **8**, 321–340 (1971)
2. Crandall, M.G., Rabinowitz, P.H.: Bifurcation, perturbation of simple eigenvalues, and linearized stability. Arch. Ration. Mech. Anal. **52**, 161–180 (1973)
3. Kanel', Ya.: On a model system of one-dimensional gas motion (Russian). Differencial'nya Uravnenija **4**, 374–380 (1968)
4. Kielhöfer, H.: Bifurcation Theory – An Introduction with Applications to PDEs. Springer, New York (2004)
5. Matsumura, A., Nishida, T.: Periodic solutions of a viscous gas equation. In: Memories, Nishida, T. (eds.) Recent Topics in Nonlinear PDE IV. Lecture Notes in Numerical and Applied Analysis, vol. 10, pp. 49–82. North-Holland, Amsterdam (1989)
6. Matsumura, A., Yanagi, S.: Uniform boundedness of the solutions for a one-dimensional isentropic model system of compressible viscous gas. Commun. Math. Phys. **175**, 259–274 (1996)
7. Weinberg S.: Gravitation and Cosmology. Wiley, New York (1972)

Computer Assisted Verification of the Eigenvalue Problem for One-Dimensional Schrödinger Operator

Ayuki Sekisaka and Shunsaku Nii

Abstract We propose a rigorous computational method for verifying the isolated eigenvalues of one-dimensional Schrödinger operator containing a periodic potential and a perturbation which decays exponentially at $\pm\infty$. We show how the original eigenvalue problem can be reformulated as the problem of finding a connecting orbit in a Lagrangian-Grassmanian. Based on the idea of the Maslov theory for Hamiltonian systems, we set up an integer-valued topological measurement, the rotation number of the orbit in the resulting one-dimensional projective space. Combining the interval arithmetic method for dynamical systems, we demonstrate a computer-assisted proof for the existence of isolated eigenvalues within the first spectral gap.

Keywords Eigenvalue problems • Connecting orbit • Topological method • Rigorous numerical

1 Introduction

We study the eigenvalue problem $Lu = \lambda u$ for the following Schrödinger operator in one dimension:

$$Lu := -u'' + q(x)u + s(x)u \tag{1}$$

where a periodic potential q with period T(i.e.,$q(x + T) = q(x)$) and a localized perturbation s are bounded C^r functions with r being sufficiently large. Note that $L : D(L) = H^2(\mathbb{R}) \subset L^2(\mathbb{R}) \to L^2(\mathbb{R})$ is a self-adjoint operator.

It is well known that the essential spectra of (1) form the band structure, which are separated by spectral gaps. They are associated with the energy gaps of electronic

A. Sekisaka (✉)
Mathematical Institute, Tohoku University, Sendai 980-8578, Japan
e-mail: sekisaka.ayuki.t4@dc.tohoku.ac.jp

S. Nii
Faculty of Mathematics, Kyushu University, Fukuoka 810-8560, Japan
e-mail: snii@math.kyushu-u.ac.jp

Y. Nishiura, M. Kotani (eds.), *Mathematical Challenges in a New Phase of Materials Science*, Springer Proceedings in Mathematics & Statistics 166,
DOI 10.1007/978-4-431-56104-0_8

states of the perfect crystal consisting of an array of atoms arranged on a periodic lattice. When an impurity is introduced, periodicity is broken and isolated energy states may or may not appear within the fundamental energy gap, depending on the strength of incorporation of impurities [5]. This semiconducting property enables us to control the electric currents in nanomaterials and its technology has offered us high-speed computers. A natural mathematical question is how the existence or non-existence of isolated eigenvalues depends on the amplitudes of the periodic and perturbation potentials. Following the reference [15], we put $q(x) = A\cos(2\pi x)$ and $s(x) = Ce^{-x^2}$ where A and C are used as given parameters. Nagatou–Plum–Nakao developed a rigorous numerical verification method, named *eigenvalue excluding* method, for proving the absence of isolated eigenvalues within the spectral gaps. They also showed numerical examples of spectral gaps applying their method to the eigenvalue problem of (1) for relatively small perturbation. For example, they obtained rigorously the interval of the first spectral gap as $[7.292924, 12.287917]$ for $(A, C) = (5, 1)$ [15].

In this paper, we propose another computational verification method based on the Maslov index theory on the Lagrangian-Grassmannian [1]. The study of eigenvalue problem (1) turns out to be equivalent to the study of one-parameter family of linear periodic Hamiltonian systems in $\mathbb{R}^2\backslash\{0\}$. We suppose that the equilibrium $\{0\}$ is hyperbolic. The fibers of its stable and unstable bundles form Lagrangian subspaces, and then the manifold of Lagrangian subspaces is called the Lagrangian-Grassmannian [3]. We reformulate the existence problem of a connecting orbit with respect to the parameter λ on this manifold whose fundamental group is $\mathbb{Z}$ (i.e., diffeomorphic to the unit circle S^1). As λ varies, the topological measurement of orbit corresponding to the rotation number of flow changes and at this change an isolated eigenvalue appears in the original problem (1).

In fact, by applying the Floquet theory to the reformulated system, we identify the band structure of essential spectra for the unperturbed case of (1). Next, introducing the polar coordinate transformation to rewrite the equation on a real projective space $\mathbb{RP}^1$, we well define the rotation index as an integer-valued topological invariant for the resulting one-dimensional dynamical systems. Applying the interval arithmetic computation to detect the points where this invariant changes with respect to λ, we obtain the accurate interval of λ containing the isolated eigenvalues of L in the presence of perturbation. Using the topological approach proposed here, we can avoid the problem of *spectral pollution* [10]. Unreal eigenvalues are caused by the break down of the usual Poincaré's min-max principle in the analytical framework approximating the linear operator. Most of the remainder of this paper will be devoted to the details of our topological method of detecting the isolated eigenvalues within the spectral gaps.

This paper is organized as follows. In Sect. 2, we trace the essential spectra for the eigenvalue problem (1) and describe the condition which must be satisfied to obtain the interval of spectral gap. In Sect. 3, we embark on our main subject of this paper, proving the main theorem that gives the rotation index of a connecting orbit. Finally, we show the computational algorithm of our topological approach, we supply a numerical example on the computer assisted proof for existence of the isolated eigenvalues within the first spectral gap for relatively large perturbation of $C \geq 6.5$.

2 Setting

This paper is concerned with one-dimensional Schrödinger operator given by Eq. (1) where $s(x)$ satisfies $s(x) \to 0$ exponentially as $|x| \to \infty$. It is well known that the spectrum of unperturbed operator:

$$L_0 u := -u'' + q(x)u \tag{2}$$

consists of only essential spectrum $\sigma_{ess}(L_0)$. In particular, it has spectral bands [12]. The essential spectrum is invariant under relatively compact perturbations and hence, the essential spectrum of L satisfies $\sigma_{ess}(L) = \sigma_{ess}(L_0) \subset \mathbb{R}$ [14, 15]. Moreover, we can restrict λ to real values because $\sigma(L) \subset \mathbb{R}$.

We are interested in the eigenvalue of L. At first we obtain the essential spectrum to restrict discussions to the eigenvalue problem. Let us consider the spectral problem:

$$L_0 u = \lambda u. \tag{3}$$

It can be rewritten as the first-order system:

$$\begin{pmatrix} u \\ v \end{pmatrix}' = A_0(x; \lambda) \begin{pmatrix} u \\ v \end{pmatrix}, \tag{4}$$

where

$$A_0(x; \lambda) = \begin{pmatrix} 0 & 1 \\ (q(x) - \lambda) & 0 \end{pmatrix} \tag{5}$$

is a T-periodic function-valued matrix depending on a real parameter λ.

Lemma 2.1 *Equation* (4) *is a periodic linear Hamiltonian system with Hamiltonian*

$$H(x, u, v; \lambda) = \frac{(\lambda - q(x))u^2}{2} + \frac{v^2}{2}. \tag{6}$$

Proof It can be easily checked that

$$\begin{aligned} A_0(x; \lambda) \begin{pmatrix} u \\ v \end{pmatrix} &= JDH(x, u, v; \lambda) \\ &= \begin{pmatrix} -\frac{\partial}{\partial v} H(x, u, v; \lambda) \\ \frac{\partial}{\partial u} H(x, u, v; \lambda) \end{pmatrix}, \end{aligned}$$

where J is a standard symplectic matrix:

$$J = \begin{pmatrix} 0 & -1 \\ 1 & 0 \end{pmatrix}.$$

□

By Floquet theory [8, 12], the evolution $\Phi(x, y; \lambda)$ of Eq. (4) is of the form

$$\Phi(x, 0; \lambda) = \Phi_{per}(x; \lambda) e^{R(\lambda)x} \tag{7}$$

where $R(\lambda) \in M_2(\mathbb{R})$ and $\Phi_{per}(x + T; \lambda) = \Phi_{per}(x; \lambda)$ for all $x \in \mathbb{R}$ with $\Phi_{per}(0; \lambda) = Id$. $\Phi(T, 0; \lambda)$ is called the Floquet or Monodoromy matrix [8]. Moreover, $\Phi(T, 0; \lambda)$ is a Hamiltonian diffeomorphism because Eq. (4) is a Hamiltonian vector field [1, 4]. In particular, $\Phi(T, 0; \lambda)$ is a symplectic and hence, it is a area-preserving transformation, that is $\det \Phi(T, 0; \lambda) = 1$ (e.g., see [4]). The matrix $\Phi(T, 0; \lambda)$ is called *hyperbolic* if it has no spectrum on the unit circle in $\mathbb{C}$. Then, we can obtain the resolvent set $\rho(L_0)$ and the essential spectrum $\sigma_{ess}(L_0)$ from the hyperbolicity of $\Phi(T, 0; \lambda)$. The resolvent set is equal to

$$\begin{aligned} \rho(L_0) &= \{\lambda \in \mathbb{R} | \sigma(R(\lambda)) \cap i\mathbb{R} = \emptyset\} \\ &= \{\lambda \in \mathbb{R} | \Phi(T, 0; \lambda) \text{ is hyperbolic}\}, \end{aligned}$$

and the essential spectrum is equal to

$$\begin{aligned} \sigma_{ess}(L_0) &= \{\lambda \in \mathbb{R} | \sigma(R(\lambda)) \cap i\mathbb{R} \neq \emptyset\} \\ &= \{\lambda \in \mathbb{R} | \Phi(T, 0; \lambda) \text{ is not hyperbolic}\}. \end{aligned}$$

Proposition 2.2

$$\begin{aligned} \sigma_{ess}(L_0) &= \{\lambda \in \mathbb{R} | \, |\operatorname{tr}(\Phi(T, 0; \lambda))| \leq 2\}, \\ \rho(L_0) &= \{\lambda \in \mathbb{R} | \, |\operatorname{tr}(\Phi(T, 0; \lambda))| > 2\}. \end{aligned}$$

Proof Since $\det(\Phi(T, 0; \lambda)) = 1$, eigenvalues $\mu_1(\lambda), \mu_2(\lambda)$ of the matrix $\Phi(T, 0; \lambda)$ satisfies $\mu_1(\lambda)\mu_2(\lambda) = 1$. Hence, if $|\operatorname{tr}(\Phi(T, 0; \lambda))| < 2$, then eigenvalues $\mu_1(\lambda)$ and $\mu_2(\lambda)$ are complex conjugate and $\mu_1(\lambda) \neq \mu_2(\lambda)$. Conversely, if $|\operatorname{tr}(\Phi(T, 0; \lambda))| > 2$, then $\mu_2(\lambda) = 1/\mu_1(\lambda)$ are satisfied. In the first case, $\mu_1(\lambda)$ and $\mu_2(\lambda)$ are on S^1 and hence $\Phi(T, 0; \lambda)$ is not hyperbolic. Therefore $\lambda \in \sigma_{ess}(L_0)$. Since $\sigma_{ess}(L_0)$ is closed in $\mathbb{R}$, we obtain the first equality. In the second case, $\mu_1(\lambda)$ and $\mu_2(\lambda)$ are real and $|\mu_1(\lambda)| > 1 > |\mu_2(\lambda)|$ holds. Therefore we obtain the second equality. □

We define stable and unstable subspaces at infinity as follows:

$$E^s_\infty(\lambda) := \{Y \in \mathbb{R}^2 | \lim_{n\to\infty} \Phi(nT, 0; \lambda)Y = 0\}, \tag{8}$$

$$E^u_\infty(\lambda) := \{Y \in \mathbb{R}^2 | \lim_{n\to-\infty} \Phi(nT, 0; \lambda)Y = 0\}. \tag{9}$$

Let $\mu_1(\lambda), \mu_2(\lambda)$ be the eigenvalues of $\Phi(T, 0; \lambda)$ with $|\mu_1(\lambda)| > 1 > |\mu_2(\lambda)|$ when $\lambda \in \mathbb{R} \setminus \sigma(L_0)$, then $E^s_\infty(\lambda)$ coincides with eigenspace associated with $\mu_1(\lambda)$ and similarly for $E^u_\infty(\lambda)$ with $\mu_2(\lambda)$.

By the continuity of eigenvalues of $\Phi(T, 0; \lambda)$, essential spectra of L_0 consists of spectral bands (or spectral bounds), that is, $\mathbb{R} \setminus \sigma(L_0) = \sqcup_i \Lambda_i$ where Λ_i are open intervals. In the next section, we characterize additional eigenvalues of L which appears under the perturbation $s(x)$ in the gap Λ_i.

3 Eigenvalues and Connecting Orbits

In this section, we restrict λ to $\mathbb{R} \setminus \sigma_{ess}(L)$ and consider the eigenvalue problem

$$Lu = -u'' + q(x)u + s(x)u = \lambda u. \tag{10}$$

It can be also rewritten by the first-order system:

$$\begin{pmatrix} u \\ v \end{pmatrix}' = A(x; \lambda) \begin{pmatrix} u \\ v \end{pmatrix}, \tag{11}$$

or simply,

$$Y' = A(x; \lambda)Y, \ Y \in \mathbb{R}^2, \tag{12}$$

where

$$A(x; \lambda) = \begin{pmatrix} 0 & 1 \\ (q(x) + s(x) - \lambda) & 0 \end{pmatrix}. \tag{13}$$

Let $\Psi(x, y; \lambda)$ be the fundamental solution matrix satisfying

$$\frac{\partial}{\partial x}\Psi(x, y; \lambda) = A(x; \lambda)\Psi(x, y; \lambda), \ \Psi(x, x; \lambda) = Id. \tag{14}$$

We define two subspaces

$$\mathbb{E}^u(y; \lambda) := \{Y \in \mathbb{R}^2 | \lim_{x\to-\infty} \Psi(x, y; \lambda)Y = 0\}, \tag{15}$$

and

$$\mathbb{E}^s(y;\lambda) := \{Y \in \mathbb{R}^2 | \lim_{x\to\infty} \Psi(x,y;\lambda)Y = 0\}. \tag{16}$$

Both subspaces define invariant vector bundles over $\mathbb{R} \times \Lambda$, where $\Lambda \subset \mathbb{R} \setminus \sigma(L_{ess})$ is the spectral gap. This means that $\mathbb{E}^{u,s}(x;\lambda) = \Psi(x,y;\lambda)\mathbb{E}^{u,s}(y;\lambda)$. Therefore, these bundles are called unstable and stable bundles. Since $\lambda \in \mathbb{R} \setminus \sigma_{ess}(L)$, and $A(x_n;\lambda) \to A_0(0;\lambda)$ as $n \to -\infty$ and $A(x_n;\lambda) \to A_0(0;\lambda)$ as $n \to \infty$ for $x_n = nT$, $\lim_{n\to-\infty} \mathbb{E}^u(x_n;\lambda) = \mathbb{E}^u_\infty(\lambda)$ and $\lim_{n\to\infty} \mathbb{E}^s(x_n;\lambda) = \mathbb{E}^s_\infty(\lambda)$. Moreover,

$$\dim \mathbb{E}^u(x;\lambda) = \dim \mathbb{E}^s(x;\lambda) = 1 \tag{17}$$

because these subspaces are invariant bundles for $\Psi(x,y;\lambda)$ and $\mathbb{E}^u_\infty(\lambda) \oplus \mathbb{E}^s_\infty(\lambda) = \mathbb{R}^2$.

We consider the eigenvalue problem on the Sobolev space $H^2(\mathbb{R})$, that is, the function space is consisting of essentially bounded functions. Therefore, the following holds.

Lemma 3.1 ([2]) *Let $\lambda \in \mathbb{R} \setminus \sigma_{ess}(L)$. Then λ is an eigenvalue of L if and only if $\mathbb{E}^u(x;\lambda) \cap \mathbb{E}^s(x;\lambda) \neq \{0\}$ for some x (and hence all $x \in \mathbb{R}$).*

By the definition, $\mathbb{E}^u(x;\lambda)$ and $\mathbb{E}^s(x;\lambda)$ satisfy $\lim_{n\to-\infty} \mathbb{E}^u(nT;\lambda) = \mathbb{E}^u_\infty(\lambda)$ and $\lim_{n\to\infty} \mathbb{E}^s(nT;\lambda) = \mathbb{E}^s_\infty(\lambda)$, respectively, where $n \in \mathbb{Z}$. By Lemma 3.1 we have the following proposition.

Proposition 3.2 *Let $\lambda \in \mathbb{R} \setminus \sigma_{ess}(L)$. Then, λ is an eigenvalue of L if and only if there exists a connecting orbit from $\mathbb{E}^u_\infty(\lambda)$ to $\mathbb{E}^s_\infty(\lambda)$ for Eq. (11).*

Proof We prove that existence of a connecting orbit is equivalent to $\mathbb{E}^u(x;\lambda) \cap \mathbb{E}^s(x;\lambda) \neq \{0\}$. If there exists a connecting orbit $Y(x;\lambda)$, then $Y(x;\lambda)$ is a bounded solution of Eq. (11) and hence, $Y(x;\lambda) \in \mathbb{E}^u(x;\lambda) \cap \mathbb{E}^s(x;\lambda)$. On the other hand, if $\mathbb{E}^u(x;\lambda) \cap \mathbb{E}^s(x;\lambda) \neq \{0\}$, then there exists a nontrivial bounded solution $Y(x;\lambda) = \Psi(x,0;\lambda)Y_0$, $Y_0 \in \mathbb{E}^u(0;\lambda) \cap \mathbb{E}^s(0;\lambda)$ such that $\psi(x;\lambda) \to 0$ as $|x| \to \infty$. Moreover,

$$\lim_{n\to-\infty} \frac{Y(nT;\lambda)}{||Y(nT;\lambda)||} \in \mathbb{E}^u_\infty(\lambda),$$

$$\lim_{n\to\infty} \frac{Y(nT;\lambda)}{||Y(nT;\lambda)||} \in \mathbb{E}^s_\infty(\lambda).$$

where $||\cdot||$ is the usual Euclidean norm of $\mathbb{R}^2$. □

Therefore, the existence problem of eigenvalues of L is rewritten as the existence problem of a connecting orbit using Proposition 3.2. We set up the topological information to obtain the information of the existence of a connecting orbit. In particular, we use the rotation number which is topological invariant and hence, we consider the existence problem of connecting orbits on the one-dimensional real projective space $\mathbb{RP}^1$, where $\mathbb{RP}^1 = \{V|\ V$ is a one-dimensional linear subspace in $\mathbb{R}^2\}$. It is well known (e.g., [3]) that the fundamental group of $\mathbb{RP}^1$ is free cyclic, that is $\pi_1(\mathbb{RP}^1) \cong \mathbb{Z}$. In particular, $\mathbb{RP}^1$ is diffeomorphic to a unit circle S^1. This means that we can define the rotation number as the topological invariants.

Let p be a natural map: $p(\alpha)$ is the equivalence class determined by α, namely $\mathrm{span}\{\alpha\} \setminus \{0\}$. If $\Psi^x = \Psi(x, 0; \lambda) : \mathbb{R}^2 \to \mathbb{R}^2$ is time-x map, then P^x, the time-x map of projectivized flow is induced by Ψ^x which the diagram

$$\begin{array}{ccc} \mathbb{R}^2 \setminus \{0\} & \xrightarrow{\Psi^x} & \mathbb{R}^2 \setminus \{0\} \\ {\scriptstyle p}\downarrow & & {\scriptstyle p}\downarrow \\ \mathbb{RP}^1 & \xrightarrow{P^x} & \mathbb{RP}^1 \end{array}$$

commutes. Therefore, Eq. (11) induces the vector field on $\mathbb{RP}^1$:

$$P' = \mathscr{P}(P, x; \lambda),\ P \in \mathbb{RP}^1. \tag{18}$$

Let $\mathscr{E}^u(\lambda) = p(\mathbb{E}^u_\infty(\lambda))$, $\mathscr{E}^s(\lambda) = p(\mathbb{E}^s_\infty(\lambda)) \in \mathbb{RP}^1$, and $P(x; \lambda)$ be the unique solution of Eq. (18) satisfying $\lim_{n\to-\infty} P(nT; \lambda) = \mathscr{E}^u_\infty(\lambda)$, that is, $P(x; \lambda) = p(\mathbb{E}^u(x; \lambda))$. Any solutions $Y(x; \lambda)$ of Eq. (11) are uniquely expressed by $Y(x; \lambda) = c_1 Y^u(x; \lambda) + c_2 Y^s(x; \lambda)$ where $Y^u(x; \lambda) \in \mathbb{E}^u(x; \lambda)$ and $Y^s(x; \lambda) \in \mathbb{E}^s(x; \lambda)$ and hence, there is only one solution such that $\lim_{n\to-\infty} P(nT; \lambda) = \mathscr{E}^u_\infty(\lambda)$.

By Proposition 3.2, the following proposition immediately holds.

Proposition 3.3 *Let $\lambda \in \mathbb{R} \setminus \sigma_{ess}(L)$. Then λ is an eigenvalue of L if and only if the solution $P(x; \lambda)$ satisfies $\lim_{n\to\infty} P(nT; \lambda) = \mathscr{E}^s(\lambda)$. Moreover, λ is not an eigenvalue of L if and only if $\lim_{n\to\infty} P(nT; \lambda) = \mathscr{E}^u(\lambda)$.*

Again we rewrite the eigenvalue problem $Lu = \lambda u$ as the existence problem of heteroclinic orbit from $\mathscr{E}^u(\lambda)$ to $\mathscr{E}^s(\lambda)$ for Eq. (18).

We introduce the useful coordinate on $\mathbb{RP}^1$ for counting the rotation number of flow of Eq. (18). Let $\theta = \tan^{-1}(\frac{v}{u})$, and we rewrite Eq. (18):

$$\theta' = (q(x) + s(x) - \lambda + 1)\cos^2\theta - 1. \tag{19}$$

Let θ be employed as the coordinates on $\mathbb{RP}^1$. Under the coordinates θ, $\mathbb{RP}^1$ is isomorphic to an interval $[-\pi/2, \pi/2]$ with the identification of endpoints. Let $\mathbb{R}$ be the universal covering of $\mathbb{RP}^1$ and regard the Eq. (19) as the equation on $\mathbb{R}$, that is, θ is regarded as an element in $\mathbb{R}$ rather than in $\mathbb{RP}^1$, and θ and $\theta + \pi$ on $\mathbb{R}$ can be identical on $\mathbb{RP}^1$.

In θ coordinate, we express $P(x; \lambda)$ with $\theta(x; \lambda)$. That is, $\theta(x; \lambda)$ is the unique solution of Eq. (19) satisfying $\lim_{n\to-\infty} \theta(nT; \lambda) = \mathscr{E}^u(\lambda)$. We define $\hat{\theta}(n; \lambda) := |\theta(nT; \lambda) - \theta(-nT; \lambda)|$. Let $\Lambda \subset \mathbb{R} \setminus \sigma_{ess}(L)$ be the spectral gap. Then, the next Theorem holds.

Theorem 3.4 *Let $[\lambda_1, \lambda_2] \subset \Lambda$ be an interval and $m \in \mathbb{N}$. If there exists $n \in \mathbb{N}$ such that $|\hat{\theta}(n; \lambda_2) - \hat{\theta}(n; \lambda_1)| > 2m\pi$, then there are at least m eigenvalues of L in $[\lambda_1, \lambda_2]$.*

Proof First assume that $|\hat{\theta}(n; \lambda_2) - \hat{\theta}(n; \lambda_1)| > 2\pi$, then, $\theta([-nT, nT]; \lambda_2)$ wrapped $\mathbb{RP}^1$ twice more than $\theta([-nT, nT]; \lambda_1)$. $\mathscr{E}^s(nT; \lambda)$ is moving in $\mathbb{RP}^1$ depends on λ, but $\mathscr{E}^s(nT; [\lambda_1, \lambda_2]) \subset \mathbb{RP}^1$. Therefore, by the continuity of a solution with respect to λ, there exists $\lambda_* \in [\lambda_1, \lambda_2]$ such that $\theta(nT; \lambda_*) = \mathscr{E}^s(nT; \lambda_*)$. Similar arguments complete our proof. □

To find the eigenvalue of L, we define an index of curve on interval $I = [\lambda_1, \lambda_2] \subset \Lambda$:

$$\mathrm{Rot}(I; n) := \left\lfloor \frac{|\hat{\theta}(n; \lambda_2) - \hat{\theta}(n; \lambda_1)|}{2\pi} \right\rfloor, \tag{20}$$

where $\lfloor \cdot \rfloor$ is the floor function. We call $\mathrm{Rot}(I; n)$ the *rotation index*. By Theorem 3.4, there exists the eigenvalues of L in $[\lambda_1, \lambda_2]$ if $\mathrm{Rot}(I; n) > 0$ for some n.

Rotation index is the useful tool for finding eigenvalues of L. However, it gives roughly information of location for the eigenvalue. Therefore, if we need more specific data of location of eigenvalue of L, we restore the number of rotations on $\mathbb{RP}^1$ to the number of intersections with the path $\theta(nT; I)$ and $\mathscr{E}^s(nT; \lambda)$ in $\mathbb{RP}^1 \times I$.

We show the following theorem for counting the intersection number. Let $\mathscr{N}_\lambda$ and $\mathscr{M}_\lambda$ be the compact neighborhood of $\mathscr{E}^u(\lambda)$ and $\mathscr{E}^s(\lambda)$ with $\mathscr{N}_\lambda \cap \mathscr{M}_\lambda = \emptyset$.

Theorem 3.5 (*Counting Intersection number*). *Let $I = [\lambda_-, \lambda_+]$ be an interval in the gap Λ and $\mathscr{M} = \cup_{\lambda \in I} \mathscr{M}_\lambda$. If n is sufficiently large, $\mathscr{M} \subset \mathbb{RP}^1$, and I has the following properties, then there exists at least one eigenvalue of L in I.*

$$\begin{aligned} \theta(nT; \lambda_-) &< \min \mathscr{M}, \\ \theta(nT; \lambda_+) &> \max \mathscr{M}. \end{aligned} \tag{21}$$

Proof By the definition of $\mathscr{E}^s(nT; \lambda)$, we can choose $n \in \mathbb{N}$ such that $\mathscr{E}^s(nT; \lambda) \subset \mathscr{M}_\lambda$ for all $\lambda \in [\lambda_-, \lambda_+]$, and we fix such number n. $\theta(nT; \lambda)$ is monotonically decreasing continuous function with respect to λ since Eq. (19) is monotone in λ, and for any λ_1, λ_2 satisfying $\lambda_1 < \lambda_2$, $\theta(nT; [\lambda_1, \lambda_2]) = [\theta(nT; \lambda_1), \theta(nT; \lambda_2)]$. Therefore, Eq. (21) means that $\mathscr{M} \subset \theta(nT; [\lambda_1, \lambda_2])$, and $\theta(nT; \lambda_-)$ and $\theta(nT; \lambda_+)$ belong to different components of $\mathbb{R} \setminus \mathscr{M}$. This implies that there exists $\lambda_* \in I$ such that $\theta(nT; \lambda_*) = \mathscr{E}^s(\lambda_*)$ and hence, this λ_* is an eigenvalue of L. □

Corollary 3.6 *Let $I = [\lambda_-, \lambda_+]$ be an interval in the gap Λ. If $\theta(nT; \lambda_-) < \min \mathscr{M}_{\lambda_-}$ and $\theta(nT; \lambda_+) > \max \mathscr{M}_{\lambda_+}$ for sufficiently large n, then there exists at least one eigenvalue of L in I.*

Proof Without loss of generality, we can assume that $\min \mathscr{M}_{\lambda_-} < \max \mathscr{M}_{\lambda_+}$. Then $\mathbb{R} \setminus [\min \mathscr{M}_{\lambda_-}, \max \mathscr{M}_{\lambda_+}]$ consists of two connected components, and $\theta(nT; \lambda_-)$

and $\theta(nT;\lambda_+)$ belongs to different components. This implies that there exists $\lambda_* \in I$ such that $\theta(nT;\lambda_*) = \mathscr{E}^s(\lambda_*)$. □

We can expect that $I = [\lambda_-, \lambda_+]$ is small because $\mathscr{E}^s(\lambda)$ is an unstable equilibrium for Eq. (19). Therefore, we may obtain a precise data about the location of eigenvalue of L. In particular, the condition of Corollary 3.6 is easily checked by numerical verification. In fact, we show that the above theorem gives us it in the next section.

We remark that these indices has a defect when apply it to the eigenvalue problem. That is, we do not obtain the non-existence or uniqueness property of the eigenvalue even if indices are equals to zero. In fact, we may obtain the information of existence of eigenvalues using Theorem 3.5 or Corollary 3.6 even if the interval I satisfies $\mathrm{Rot}(I;n) = 0$, and our results express nothing about information of non-existence and uniqueness of eigenvalues.

4 Sample Result

The argument in previous sections is arranged into the following 3 steps.

Step 1. Construction of an interval $[\lambda_-, \lambda_+]$ in the spectral gap.
Step 2. Determination of an integral interval $[-nT, nT]$ and enclosure of $\mathscr{E}^u(\lambda), \mathscr{E}^s(\lambda)$ for the Eq. (4).
Step 3. Enclosure of a heteroclinic orbit $\theta(nT;\lambda)$ from $\mathscr{E}^u(\lambda)$ to $\mathscr{E}^s(\lambda)$.

In this section, we verify eigenvalues of L by the above method for the case

$$L_C u = -u'' + 5\cos(2\pi x)u + Ce^{-x^2}u \tag{22}$$

where $C \in \mathbb{R}$ is a parameter. This example (22) is also discussed in the paper [15]. In this case, $q(x) = 5\cos(2\pi x)$ and $s(x) = Ce^{-x^2}$ and hence, a period T of periodic potential is equal to 1. We use of the software package CAPD (ver 2.0) [9] in step 2 and 3 to obtain the rigorous result in the strict mathematical sense. CAPD library is a collection of flexible C++ modules which are mainly designed to computation of homology of sets, and validated numerics for dynamical systems.

Step 1. We compute the Floquet matrix $\Phi(1,0;\lambda)$ for the following equation:

$$\begin{pmatrix} u \\ v \end{pmatrix}' = \begin{pmatrix} 0 & 1 \\ 5\cos(2\pi x) - \lambda & 0 \end{pmatrix} \begin{pmatrix} u \\ v \end{pmatrix}. \tag{23}$$

By Proposition 2.2, approximate spectral gaps are given in Table 1. In this step, we do not need validated numerics for the spectral gap. Then, in the following discussions, we restrict a parameter λ in [7.3, 12.2].

Step 2. Let $\mathbb{E}^s(x;\lambda)$ and $\mathbb{E}^u(x;\lambda)$ be the stable and unstable subspaces satisfying $\lim_{n\to\infty} \mathbb{E}^s(n;\lambda) = \mathbb{E}^s_\infty(\lambda)$ and $\lim_{n\to-\infty} \mathbb{E}^u(n;\lambda) = \mathbb{E}^u_\infty(\lambda)$, respectively. We

Table 1 Spectral gap

λ	$\lvert \mathrm{tr}\, \Phi(1, 0; \lambda)\rvert$
7.29292	1.999999386140228 < 2
7.29293	2.000000827693733 > 2
12.28791	2.000000864383513 > 2
12.28792	1.999999744929798 < 2

Table 2 Enclosure of the stable fixed point and time-1 map for the case $C = 6.5, n = 9$

λ	$N_1(\lambda)$	$\mathscr{F}(N_1; \lambda)$
7.3	[1.5572146772, 1.5572146792]	[1.55721467726, 1.55721467915]
7.4	[1.5178189939, 1.5178189959]	[1.51781899411, 1.51781899572]
7.5	[1.4969022731, 1.4969022751]	[1.49690227334, 1.49690227483]
7.6	[1.4805161372, 1.4805161392]	[1.4805161375, 1.48051613891]
⋮	⋮	⋮
8.0	[1.4315770104, 1.4315770124]	[1.43157701083, 1.43157701205]
⋮	⋮	⋮
12.2	[0.444472223, 0.444472225]	[0.444472223076, 0.444472224935]

must obtain the integral interval $[-n, n]$, and enclosures $p(\mathbb{E}^s(n; \lambda)) \in [\mathscr{E}^s(\lambda)]$ and $p(\mathbb{E}^u(n; \lambda)) \in [\mathscr{E}^u(\lambda)]$ where $\mathscr{E}^{s,u}(\lambda) = p(\mathbb{E}^{s,u}_{\infty}(\lambda))$, respectively. Therefore, we use the differential inclusion:

$$\theta' \in (5\cos(2\pi x) + [\delta] - \lambda + 1)\cos^2\theta - 1, \tag{24}$$

where $[\delta]$ is an interval $[0, Ce^{-x^2}]$. Let $\mathscr{F}(N; \lambda)$ be enclosure of solution of Eq. (24) from 0 to 1 with the initial data N. If set N satisfies $\mathscr{F}(N; \lambda) \subseteq N$ in $\mathbb{RP}^1$, then N contains the unique fixed point for time 1-map of Eq. (24). Hence, we should just verify that $\mathscr{F}(N_i(\lambda); \lambda) \subset N_i(\lambda)$, $i = 1, 2$, where $N_1(\lambda) = [\mathscr{E}^u(\lambda)]$ and $N_2(\lambda) = [\mathscr{E}^s(\lambda)]$ are enclosure of each fixed point in $\mathbb{RP}^1$ satisfying $N_1(\lambda) \cap N_2(\lambda) = \emptyset$. Remark that if the interval $N(\lambda)$ is a neighborhood of unstable fixed point for Eq. (24) then $-N(\lambda)$ is the neighborhood of stable fixed point for Eq. (24). It means that unstable fixed point $\mathscr{E}^u(\lambda)$ and stable fixed point $\mathscr{E}^s(\lambda)$ satisfies $\mathscr{E}^s(\lambda) = -\mathscr{E}^u(\lambda)$ on θ coordinates in this model. Therefore, $|\mathscr{E}^{u,s}(\Lambda)| < \frac{\pi}{2}$. In fact, the enclosure of $\mathscr{E}^u(\lambda)$ is moving in the range $[0, \frac{\pi}{2}]$ depending λ in the first gap [7.3, 12.2] (see Table 2).
Then we obtain the pair of $N_1(\lambda)$ and $N_2(\lambda)$ satisfying $N_1(\lambda) \cap N_2(\lambda) = \emptyset$ and they have a fixed point as in Table 2. Note that (24) does not have fixed points when $\lambda \in \sigma_{ess}(L)$. Therefore, it implies that $\lambda \in \mathbb{R} \setminus \sigma_{ess}(L)$.
Step 3. We consider the equation

$$\theta' \in (5\cos(2\pi x) + Ce^{-x^2} - \lambda + 1)\cos^2\theta - 1, \tag{25}$$

and show that there exists a parameter λ such that there exists a connecting orbit from $\mathscr{E}^u(\lambda)$ to $\mathscr{E}^s(\lambda)$. Using the rotation index $\mathrm{Rot}(I;\lambda)$, there exists the connecting orbit for the specific λ in $[7.4, 8.0]$. Then, we can replace the definition of rotation number $\mathrm{Rot}(I;\lambda)$ by the half rotation number:

$$\mathrm{Rot}(I;n) := \left\lfloor \frac{|\hat{\theta}(n;\lambda_2) - \hat{\theta}(n;\lambda_1)|}{\pi} \right\rfloor,$$

using the fact which is proved in **step 2**.

Therefore, by Theorem 3.4 or Corollary 3.6, at least one eigenvalue exists in the interval $[7.4, 8.0]$. In particular, we can easily check the condition of Corollary 3.6.

If the result of the range for eigenvalue existence is not enough, then we can use Theorem 3.5 to raising accuracy. We give the following subroutine algorithm for obtaining more accurate data.

We choose λ satisfying $7.4 < \lambda < 8.0$. Then we obtain new intervals $\Lambda_1 = [7.4, \lambda]$ and $\Lambda_2 = [\lambda, 8]$, and we go back to Step 1. If each new interval $\Lambda_i,\ i = 1, 2$ satisfy $Rot(\Lambda_i; 9) = 0$, then we may obtain the eigenvalue in $\Lambda_i,\ i = 1, 2$ using Theorem 3.5. In fact, we get smaller interval in $[7.4, 7.6]$ than interval $[7.4, 8.0]$ containing eigenvalues of L. If we want to get smaller interval, repeat this step until you obtain results with a desired accuracy. For example, we can get the following result.

Computer Assisted Result 4.1 *When $C = 7.0$, there is at least one eigenvalue of L in the interval* $[7.6151, 7.6160]$.

Nagatou [15] showed that, in case of $C = 1$, there exists an eigenvalue-free interval $[7.55, 11.335]$. Therefore, we may think that the eigenvalue in $[7.4, 7.6]$ with $C = 6.5$ generated when C is between 1.0 and 6.5. Moreover, we are interested in the asymptotic behavior of eigenvalues in a given subinterval of the gap as the constant C tends to infinity. A variety of studies is concerned with such behavior (e.g. Deift–Hempel [11], Hempel [13]).

Let $\mathbb{F}$ be the set of floating point numbers and $\mathbb{IF}$ be the set of intervals whose end-points are in $\mathbb{F}$. Similarly, $\mathbb{IF}^n$ is the set of n-dimensional cubes, that is,

$$\mathbb{IF}^n := \{I_1 \times I_2 \times \cdots \times I_n | I_i \in \mathbb{IF}\}. \tag{26}$$

We define the parameter space for Eq. (24) as $(\lambda, C) \in \Lambda \times \mathscr{C} = \cup_i \Lambda_i \times \cup_i \mathscr{C}_i \subset \mathbb{IF}^2$ where $\Lambda_i = [\lambda_i, \lambda_{i+1}]$ and $\mathscr{C}_i = [C_i, C_{i+1}]$. First, using our method, we obtain intervals $\mathscr{I}_C$ containing eigenvalues for each fixed C. Next, for any j, we check off $\mathscr{Q}_{k\ell} = \Lambda_k \times \mathscr{C}_\ell$ if $\Lambda_k \times \{C_\ell\}$ or $\Lambda_k \times \{C_{\ell+1}\}$ contains eigenvalues. By the above procedures, we obtain enclosures of eigenvalue branches in $\mathscr{C} \times \Lambda$ as following.

Computer Assisted Result 4.2 *There exists at least one pair (λ, C) in each red rectangle $\mathscr{Q}$ in* $[7.3, 12.2] \times [6.0, 30.0]$ *in Fig. 1, such that the connecting orbit from $\mathscr{E}^s(\lambda)$ to $\mathscr{E}^u(\lambda)$ exists. Therefore, at least one eigenvalue of L_C exists for each parameter C.*

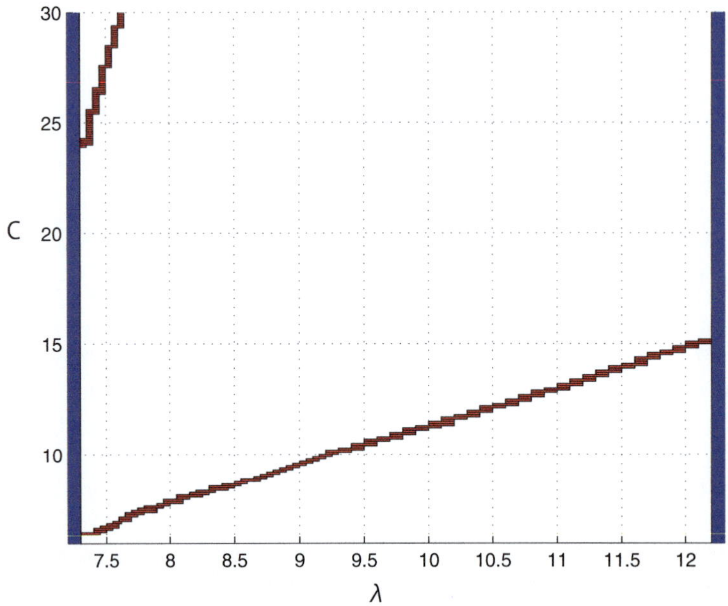

Fig. 1 Distribution of the eingenvalue: The *horizontal line* means λ and the *vertical line* means the parameter C. Each *red-colored rectangle* (*shaded regions*) in $\Lambda \times \mathscr{C} = [7.3, 12.2] \times [6.0, 30.0]$ contains at least one pair (λ, C) where there is the connecting orbit from $\mathscr{E}^s(\lambda)$ to $\mathscr{E}^u(\lambda)$. Therefore, either the upper or lower or both edge of each *rectangle* has at least one eigenvalue of L_C for each parameter C. Outer region (*blue-colored region*) of region $\Lambda \times \mathscr{C}$ contains spectral gaps

5 Discussions

We have proposed a rigorous computational method for verifying the isolated eigenvalues for the one-dimensional Schrödinger operator containing a periodic potential and a perturbation which decays exponentially at $\pm\infty$. Combining the interval arithmetic method for dynamical systems, we have demonstrated a computer-assisted proof for the existence of isolated eigenvalues. Reformulating the original eigenvalue problem to the problem of a connecting orbit for the first order ODE on the unit circle, we can substantially reduce the amount of computational effort to obtain the global behavior of isolated eigenvalues within the spectral gap. Actually, our computations are performed on an Apple MacPro (2.53 GHz Intel Core 2 CPU) computer. It required a few minutes to obtain each interval indicated as a red-colored rectangle in Fig. 1.

Based on the idea of the Maslov theory for Hamiltonian systems, we set up explicitly the integer-valued topological invariants, the rotation and intersection numbers of the orbit in the one-dimensional projective space, which are associated with the information of isolated eigenvalue and its associated eigenfunction profile, respectively. It is well known that topological invariants like rotation index we proposed here are useful tools for studying the instabilities of the solutions in nonlinear differential equations. However, setting up a practical way to count the index seems to be difficult. Finally, it is worth noting that Chardard et al. [6, 7] have also developed the computational method using the Maslov index to investigate the stability of solitary waves in higher-dimensional systems. They used a shooting argument to show the existence of eigenvalues. Applications of our computer-assisted proof method to the realistic problems including the energy band structures of three-dimensional photonic crystal remains a challenge for the future.

Acknowledgements We would like to thank Professor Kaori Nagatou for the helpful suggestions. Special thanks go to Professor Yasumasa Nishiura and Takashi Teramoto for many valuable comments, and we are also grateful to the referees for suggesting several improvements.

References

1. Abbondandolo, A.: Morse Theory for Hamiltonian Systems. Chapman Hall/CRC, Boca Raton (2001)
2. Alexander, J.W., Gardner, R., Jones, C.K.R.T.: A topological invariant arising in the stability analysis of traveling waves. J. Reine Angew. Math. **410**, 167–212 (1990)
3. Arnol'd, VI.: Characteristic class entering in quantization conditions. Funct. Anal. Appl. **1**, 1–13 (1967)
4. Arnol'd, VI.: Mathematical Methods of Classical Mechanics, 2nd edn. Springer, New York (1989)
5. Ashcroft, N.W., Mermin, N.D.: Solid State Physics. HRW, New York (1976)
6. Chardard, F., Dias, F., Bridges, T.J.: Fast computation of the Maslov index for hyperbolic periodic orbits. J. Phys. A: Math. Gen. **39**, 14545–14557 (2006)
7. Chardard, F., Dias, F., Bridges, T.J.: Computing the Maslov index of solitary waves. Part 1: Hamiltonian systems on a 4-dimensional phase space. Physica D **238**, 1841–1867 (2010)
8. Coddington, E.A., Levinson N.: Theory of Ordinary Differential Equations, vol. XII. McGill-Hill, New York/Toront/London (1955)
9. Computer Assisted Proofs in Dynamics Group (CAPD) (website). http://capd.ii.uj.edu.pl/index.php
10. Davies, E., Plum, M.: Spectral pollution. IMA J. Numer. Anal. **24**, 417–438 (2004)
11. Deift, P.A., Hempel, R.: On the existence of eigenvalues of the Schrödinger operator $H - \lambda W$ in a gap of $\sigma(H)$. Commun. Math. Phys. **103**, 461–490 (1986). doi:10.1006/jfan.1999.3542
12. Eastham, M.S.P.: The Spectral Theory of Periodic Differential Equations. Scottish Achademic Press, Edinburgh (1973)
13. Hempel, R.: On the asymptotic distribution of the eigenvalue branches of a Shrödinger operator $H - \lambda W$ in a spectral gap of H. J. Reine Angew. Math. **399**, 38–59 (1989)
14. Kato, T.: Perturbation Theory for Linear Operators. Reprint of the 1980 edition, Springer, Berlin/Heidelberg (1995)
15. Nagatou, K., Plum, M., Nakao, M.: Eigenvalue excluding for perturbed-periodic one-dimensional Schrödinger operators. Proc. R. Soc. A **159** (2011). doi:10.1098/rspa.2011.0159

MIX
Papier aus verantwortungsvollen Quellen
Paper from responsible sources
FSC® C105338

If you have any concerns about our products,
you can contact us on
ProductSafety@springernature.com

In case Publisher is established outside the EU,
the EU authorized representative is:
Springer Nature Customer Service Center GmbH
Europaplatz 3, 69115 Heidelberg, Germany

Printed by Libri Plureos GmbH
in Hamburg, Germany